Reviews of Environmental Contamination and Toxicology

VOLUME 186

T0137760

Reviews of Environmental Contamination and Toxicology

Continuation of Residue Reviews

Editor
George W. Ware

VOLUME 186

 Springer

Springer
New York: 233 Spring Street, New York, NY 10013, USA
Heidelberg: Postfach 10 52 80, 69042 Heidelberg, Germany

ISBN 978-1-4419-2114-7   e-ISBN 978-0-387-32883-6
ISSN 0179-5953

Printed on acid-free paper.

springer.com

Foreword

International concern in scientific, industrial, and governmental communities over traces of xenobiotics in foods and in both abiotic and biotic environments has justified the present triumvirate of specialized publications in this field: comprehensive reviews, rapidly published research papers and progress reports, and archival documentations. These three international publications are integrated and scheduled to provide the coherency essential for nonduplicative and current progress in a field as dynamic and complex as environmental contamination and toxicology. This series is reserved exclusively for the diversified literature on "toxic" chemicals in our food, our feeds, our homes, recreational and working surroundings, our domestic animals, our wildlife and ourselves. Tremendous efforts worldwide have been mobilized to evaluate the nature, presence, magnitude, fate, and toxicology of the chemicals loosed upon the earth. Among the sequelae of this broad new emphasis is an undeniable need for an articulated set of authoritative publications, where one can find the latest important world literature produced by these emerging areas of science together with documentation of pertinent ancillary legislation.

Research directors and legislative or administrative advisers do not have the time to scan the escalating number of technical publications that may contain articles important to current responsibility. Rather, these individuals need the background provided by detailed reviews and the assurance that the latest information is made available to them, all with minimal literature searching. Similarly, the scientist assigned or attracted to a new problem is required to glean all literature pertinent to the task, to publish new developments or important new experimental details quickly, to inform others of findings that might alter their own efforts, and eventually to publish all his/her supporting data and conclusions for archival purposes.

In the fields of environmental contamination and toxicology, the sum of these concerns and responsibilities is decisively addressed by the uniform, encompassing, and timely publication format of the Springer triumvirate:

Reviews of Environmental Contamination and Toxicology [Vol. 1 through 97 (1962–1986) as Residue Reviews] for detailed review articles concerned with any aspects of chemical contaminants, including pesticides, in the total environment with toxicological considerations and consequences.

Bulletin of Environmental Contamination and Toxicology (Vol. 1 in 1966)
for rapid publication of short reports of significant advances and dis-
coveries in the fields of air, soil, water, and food contamination and pollution as well as methodology and other disciplines concerned with the
introduction, presence, and effects of toxicants in the total environment.

Archives of Environmental Contamination and Toxicology (Vol. 1 in 1973)
for important complete articles emphasizing and describing original
experimental or theoretical research work pertaining to the scientific
aspects of chemical contaminants in the environment.

Manuscripts for *Reviews* and the *Archives* are in identical formats
and are peer reviewed by scientists in the field for adequacy and value;
manuscripts for the *Bulletin* are also reviewed, but are published by photo-
offset from camera-ready copy to provide the latest results with minimum
delay. The individual editors of these three publications comprise the joint
Coordinating Board of Editors with referral within the Board of manuscripts submitted to one publication but deemed by major emphasis or
length more suitable for one of the others.

Coordinating Board of Editors

Preface

The role of *Reviews* is to publish detailed scientific review articles on all aspects of environmental contamination and associated toxicological consequences. Such articles facilitate the often-complex task of accessing and interpreting cogent scientific data within the confines of one or more closely related research fields.

In the nearly 50 years since *Reviews of Environmental Contamination and Toxicology* (formerly *Residue Reviews*) was first published, the number, scope and complexity of environmental pollution incidents have grown unabated. During this entire period, the emphasis has been on publishing articles that address the presence and toxicity of environmental contaminants. New research is published each year on a myriad of environmental pollution issues facing peoples worldwide. This fact, and the routine discovery and reporting of new environmental contamination cases, creates an increasingly important function for *Reviews*.

The staggering volume of scientific literature demands remedy by which data can be synthesized and made available to readers in an abridged form. *Reviews* addresses this need and provides detailed reviews worldwide to key scientists and science or policy administrators, whether employed by government, universities or the private sector.

There is a panoply of environmental issues and concerns on which many scientists have focused their research in past years. The scope of this list is quite broad, encompassing environmental events globally that affect marine and terrestrial ecosystems; biotic and abiotic environments; impacts on plants, humans and wildlife; and pollutants, both chemical and radioactive; as well as the ravages of environmental disease in virtually all environmental media (soil, water, air). New or enhanced safety and environmental concerns have emerged in the last decade to be added to incidents covered by the media, studied by scientists, and addressed by governmental and private institutions. Among these are events so striking that they are creating a paradigm shift. Two in particular are at the center of ever-increasing media as well as scientific attention: bioterrorism and global warming. Unfortunately, these very worrisome issues are now super-imposed on the already extensive list of ongoing environmental challenges.

The ultimate role of publishing scientific research is to enhance understanding of the environment in ways that allow the public to be better informed. The term "informed public" as used by Thomas Jefferson in the age of enlightenment conveyed the thought of soundness and good judgment. In the modern sense, being "well informed" has the narrower meaning of having access to sufficient information. Because the public still gets most of its information on science and technology from TV news and reports, the role for scientists as interpreters and brokers of scientific information to the public will grow rather than diminish.

Environmentalism is the newest global political force, resulting in the emergence of multi-national consortia to control pollution and the evolution of the environmental ethic. Will the new politics of the 21st century involve a consortium of technologists and environmentalists, or a progressive confrontation? These matters are of genuine concern to governmental agencies and legislative bodies around the world.

For those who make the decisions about how our planet is managed, there is an ongoing need for continual surveillance and intelligent controls, to avoid endangering the environment, public health, and wildlife. Ensuring safety-in-use of the many chemicals involved in our highly industrialized culture is a dynamic challenge, for the old, established materials are continually being displaced by newly developed molecules more acceptable to federal and state regulatory agencies, public health officials, and environmentalists.

Reviews publishes synoptic articles designed to treat the presence, fate, and, if possible, the safety of xenobiotics in any segment of the environment. These reviews can either be general or specific, but properly lie in the domains of analytical chemistry and its methodology, biochemistry, human and animal medicine, legislation, pharmacology, physiology, toxicology and regulation. Certain affairs in food technology concerned specifically with pesticide and other food-additive problems may also be appropriate.

Because manuscripts are published in the order in which they are received in final form, it may seem that some important aspects have been neglected at times. However, these apparent omissions are recognized, and pertinent manuscripts are likely in preparation or planned. The field is so very large and the interests in it are so varied that the Editor and the Editorial Board earnestly solicit authors and suggestions of underrepresented topics to make this international book series yet more useful and worthwhile.

Justification for the preparation of any review for this book series is that it deals with some aspect of the many real problems arising from the presence of foreign chemicals in our surroundings. Thus, manuscripts may encompass case studies from any country. Food additives, including pesticides, or their metabolites that may persist into human food and animal feeds are within this scope. Additionally, chemical contamination in any manner of air, water, soil, or plant or animal life is within these objectives and their purview.

Manuscripts are often contributed by invitation. However, nominations for new topics or topics in areas that are rapidly advancing are welcome. Preliminary communication with the Editor is recommended before volunteered review manuscripts are submitted.

Tucson, Arizona   G.W.W.

Table of Contents

Rev Environ Contam Toxicol 186:1–56 © Springer 2006

Health Risks of Enteric Viral Infections in Children

Nena Nwachuku and Charles P. Gerba

Contents

Communicated by Charles P. Gerba

N. Nwachuku (✉)
Office of Science and Technology, Office of Water, U.S. Environmental Protection Agency, 1200 Pennsylvania Ave. N.W., Mail Code 4304T, Washington, D.C. 20460, U.S.A.

C.P. Gerba
Department of Soil, Water and Environmental Science, University of Arizona, Tucson, AZ 85721, U.S.A.

I. Introduction

A growing body of scientific knowledge demonstrates that children (persons less than 18 yr of age) may suffer disproportionately from some environmental risks. These risks may arise because children's neurological, immunological, and digestive systems are still developing. In addition, children are more exposed to pathogens in the environment because of poor or lack of sanitary habits. Because all enteric microorganisms have a potential to be transmitted by the fecal–oral route, waterborne exposure is a major concern.

Children are potentially at a greater risk of infections from serious enteric viral illness for a number of reasons (Table 1). Most important is the immune system, which is needed to control the infection processes; this can lead to more serious infections than in adults, who have fully developed immune systems.

It has been shown that the immune system in the infant is immature. B-lymphocyte function is immature in neonates and impaired in children less than 2 yr of age, which is partially caused by immature T-cell helper function (Cummins et al. 1994). The spleen in children <2 yr of age is characterized by an immature marginal zone compartment, which indicates that B cells are not as well developed and do not react to antigenic stimuli as effectively (Timens et al. 1989).

Immunoglobulin A (IgA) is essential in combating microbial infections in the gastrointestinal tract. Levels increase with age, with adults having 70–300 times that of newborns (Roy 1995). Even at 10 yr of age, IgA levels are half those found in an adult. It has been demonstrated that there is an increase in macrophage production, and the percentage of these cells expressing immunoreactive interferon-alpha in infant lungs is lower when compared with fetal lungs (Khan et al. 1990).

Heinberg et al. (1964) proposed that infection with coxsackievirus B1 in young mice is fatal because of an inadequate production of interferon, whereas older animals that produce interferon can usually survive these infections. It is thought that foreign nucleic acids must enter the cell for interferon to be produced, but it has also been proposed that viral infec-

Table 1. Why Children are at Greater Risk of Gastrointestinal Infections.

Immature immune system
Intestinal mucosa more permeable to water
Proportionately less extracellular fluid than adults
Physiological deficiency in IgA[a]
Reduced stomach acid and pepsin secretion
Absent or poor sanitation habits

[a]Immunoglobulin A, a group of antibodies in bodily secretions.

tion can occur so quickly that cells are destroyed before they can produce interferon (Kunin 1964). In a study evaluating the increased susceptibility of newborns to infections by group B streptococci, it was proposed that phagocytic defenses *in utero* and at birth may be lacking. The newborn is unable to produce enough phagocytes in adequate numbers to deliver them to the site of infection (Wilson 1986); this may also play a role in enteroviral infections. Lung macrophages from monkeys and rabbits less than 7 d old have less microbiocidal activity than lung macrophages from adults (Bellanti et al. 1979; Jacobs et al. 1983).

The digestive system of children is also not fully developed, resulting in greater susceptibility to enteric viral disease. Enteric pathogens need to pass through the harsh acidic environment of the stomach before they can initiate infection and disease. Infants and undernourished children may have a reduced acid and pepsin secretion by the gastric mucosa; as a result, these agents may better survive the transit from the stomach to the intestine.

The intestinal tract of the newborn mouse has been shown to be an ineffective barrier against orally acquired group B coxsackieviruses, whereas the adult mouse is very resistant to infection by this agent (Loria et al. 1976). The intestinal tract of the adult mouse offers two modes of protection against oral infection. First, it acts as a barrier that prevents the virus from passing through the mucosal side of the gut and, second, provides a clearance mechanism that eliminates the virus from the enteric tract after infection (Loria et al. 1976).

Infants and children are more susceptible to dehydration because secreted liquids, as a proportion of extracellular fluid, may be twice as much as that of adults. Infants are at higher risk of dehydration, as their intestinal mucosa tends to be more permeable to water. Therefore, the same pathologic process in infants may result in greater loss of water and electrolytes than in older children whose intestinal mucosa is less permeable to water (Roy 1995).

The neurological system of children may also be less likely to defend against a viral infection. Enterovirus infections are generally more severe in infant humans and mice. It has been shown that the brains of mice less than 48 hr old are more prone to infection by coxsackievirus B and typically die of encephalitis, whereas the brains of adult mice are resistant to infection (Khatib et al. 1980).

The heart and endocrine system are also more likely to be affected in children than in adults during enterovirus infection. Myocarditis is typical among mice less than 48 hr old infected with coxackievirus B1 and B4, but infected mice less than 14 d old experience myocarditis less frequently. Adult mice infected with coxsackievirus B1 or B4 do not develop myocarditis (Khatib et al. 1980). Pancreatitis is a more common condition in mice less than 14 d old than in adult mice.

Exposure to fecal–oral agents is also more likely in children. Infants are not yet developmentally capable of habits that would reduce their exposure; these include toilet use and handwashing. The frequency of hand-to-mouth or fomite (inanimate object)-to-mouth contact is also greater in children. Small children may bring their hands to their mouths once every 3 min (Springthorpe and Sattar 1990).

II. Viral Diseases in Infancy and Childhood
A. Diarrheal Diseases

Diarrheal diseases are the leading cause of childhood morbidity and mortality in developing countries (Kosek et al. 2003). It has been estimated (LeBaron et al. 1990) that infectious gastroenteritis in the United States alone causes more than 210,000 children of 5 yr of age or younger to be hospitalized, at a yearly cost of nearly $1 billion. Throughout the world, 3–5 billion cases of diarrhea occur (LeBaron et al. 1990). In the U.S., 250–350 million cases of diarrhea occur every year, with more than 4,000 deaths in all age groups (Glass 2000).

Worldwide, more than 500 million episodes of diarrhea occur each year in children under 5 yr of age (Fimberg et al. 1993), with an associated 2.5 million deaths, representing the leading cause of infant mortality. Approximately half of those deaths are among children under 1 yr of age (Kosek et al. 2003). Case fatality rates average from 1.8% in children under 1 yr of age to 0.15% in children aged under 5 yr (Kosek et al. 2003). Although developing countries account for the majority of all diarrheal disease deaths, the diarrheal syndrome is one of the 10 leading causes of death in infants in the U.S. An estimated 21–37 million episodes of diarrhea occur annually in children under 5 yr of age; 10% of the affected children are seen by a physician, more than 200,000 are hospitalized, and 300–400 die of the illness (Wyllie 1999).

From 1968 to 1991, a total of 14,137 deaths associated with diarrhea were reported in the U.S. Infants (1–11 mon) accounted for 78% of these deaths. Although the median age at the time of death, from 1961 to 1985, was reduced from 5 to 1.5 mon and the disease mortality decreased by 75%, no further reduction in mortality has been recorded since then, with a yearly average of 300 gastroenteritis-associated deaths (Kilgore et al. 1995).

No etiological agent is identified in the majority (79.5%) of hospital admissions of children associated with gastroenteritis in the U.S. However, viral gastroenteritis is the leading cause of identified causes of diarrhea-associated hospitalization in children, accounting for 25.3% of such admissions, followed by bacteria (5.4%) and parasites (<0.3%) (Parashar et al. 1999). Rotavirus, adenovirus, Norwalk virus, astrovirus, and calicivirus are the most common viral pathogens, rotavirus being the most common, with 16.5% of all diarrhea-associated hospital admissions (Newman et al. 1999). In Connecticut, the annual incidence of diarrhea-associated hospitalizations from 1987 through 1996 was 49.4 per 10,000, whereas the cumulative incidence over the first 5 yr of life was 247 per 10,000. A breakdown by ethnicity and race indicates that the incidence of diarrhea-associated hospitalizations is greater in Hispanics and African-Americans than in Anglo-whites (Parashar et al. 1999).

Greater Susceptibility of Children. Age plays a significant role in the pathogenesis of diarrhea. It was shown in the Philippines that age was inversely related to the duration of diarrhea in children (San Pedro and Waltz 1991). Many of the host defense mechanisms of the enteric mucosa are immature or inefficient in the newborn. Secretory immunoglobulin A (sIgA) provides a major line of defense against enteric pathogens. However, in infants and children there is a physiological deficiency of this type of antibody. This deficiency is partially compensated by breastfeeding (Roy 1995).

There is indirect evidence that some, still unidentified, components of maternal milk may enhance the ability of the infant to mount a more vigorous immune response (Haffejee et al. 1990). Breast-fed infants had significantly higher antibody titers (i.e., concentration) to the oral polio vaccine (OPV) than infants fed formulas (Pickering et al. 1998). A study conducted in Sri Lanka showed that infants breast-fed for 3 mon or less developed respiratory and gastroentric infections earlier than infants who were breast-fed for longer periods, whereas those who were never breast-fed suffered earlier and more severe infections (Perera et al. 1999). In addition, in the first year of life, the incidence of diarrheal illnesses among breast-fed infants is half that of formula-fed infants (Dewey et al. 1995).

Infants and children are more susceptible to dehydration because the absorbed and secreted liquids, as a proportion of extracellular fluid, may be as much as twice that of adults. Infants are at highest risk of dehydration because their intestinal mucosa tends to be more permeable to water. Therefore, the same pathological process in infants may result in a greater loss of water and electrolytes than in older children whose intestinal mucosa is less permeable to water (Roy 1995).

Enteric pathogens need to pass through the harsh acidic environment of the stomach before they can initiate infection and disease. Infants and undernourished children may have reduced acid and pepsin secretion by the gastric mucosa; as a result, these agents may survive better the transit from the stomach to the intestine. Furthermore, undernourished infants have a reduced ability to replace lost epithelial cells of the intestinal mucosa (Sherman and Litchman 1995). Ironically, it has been postulated that the relatively uncommon occurrence of rotavirus gastroenteritis in neonates may be related, in part, to their immature proteolytic enzyme inability to cleave viral protein 4 (VP4) of virulent strains, rendering them noninfectious (Haffejee 1995).

Low birth weight has been identified as a significant risk factor for infant hospitalization with viral gastroenteritis. Infants weighing less than 1.5 kg were at highest hospitalization risk, whereas those weighing more than 4.0 kg had a decreased risk (Newman et al. 1999).

In addition to physiological differences between children and adults, young children seldom follow proper sanitary practices unless closely supervised. Therefore, children are more prone to acquire infections transmitted through the fecal–oral route (Sherman and Litchman 1995).

Rotavirus. Rotaviruses are the most important agents of infantile gastroenteritis around the world. The antigenic characteristics of rotaviruses are defined by group, subgroup, and serotype. There are seven rotavirus groups (A through G), whose specificity is given by the viral protein VP6. All seven groups are found in animals, but only A, B, and C are found in humans (Kapikian 1996).

Group A rotavirus is endemic worldwide. It is the leading cause of severe diarrhea among infants and children, and accounts for about half of the cases requiring hospitalization. In temperate areas, it occurs primarily in the winter (Kapikian and Chanock 1996), but in the tropics it may occur throughout the year (San Pedro and Waltz 1991; Stewien et al. 1991; Kapikian 1996). Worldwide rotavirus infections in children cause 2 million hospitalizations, and 352,000–592,000 deaths in children less than 5 yr of age (Parasher et al. 2003). In the U.S. alone, as many as 18,000 hospitalizations occur each year from rotavirus (Fischer et al. 2004).

Group B rotavirus, also called adult diarrhea rotavirus, has caused major epidemics of severe diarrhea affecting thousands of persons of all ages in China (Ramachandran et al. 1998). Group C rotavirus has been associated with rare and sporadic cases of diarrhea in children in many countries (Kapikian and Chanock 1996). Rotavirus infections are very common in both developed and underdeveloped countries, as evidenced by the prevalence of serum antibodies, which are found in the majority of children by 3 yr of age.

In adults, rotaviral infection is usually subclinical. However, rotavirus gastroenteritis outbreaks have occurred in army recruits, geriatric patients, and hospital staff (Kapikian and Chanock 1996). Over 3 million cases of rotavirus gastroenteritis occur annually in the U.S. It has been estimated that rotaviruses, in the U.S. alone, cause more than 1 million cases of severe diarrhea and up to 150 deaths per year. Worldwide, close to 1 million infants and young children die of rotavirus infection each year. Rotavirus infection does not result in an efficient or long-lasting immunity. Therefore, rotavirus infection in the same child often occurs up to six times during childhood (Kapikian and Chanock 1996).

Most reports agree that the highest incidence of rotavirus gastroenteritis occurs in children 6–24 mon old; however, some studies have found the highest incidence at 6–12, and 9–14 mon. In general, infants in developing countries tend to become infected by rotavirus much earlier than children in the developed world (Haffejee 1995).

Apparently, rotavirus infection occurs at a very early age. A study conducted in South Africa showed that 15 of 19 rotavirus-infected asymptomatic neonates shed rotavirus during the first 5 d of age; among those, 2 excreted the virus during the first 24 hr of life (Haffejee et al. 1990). Nevertheless, rotavirus gastroenteritis in neonates is relatively uncommon and when present is usually mild. It has been established that approximately 80%–90% of rotavirus-infected babies remain asymptomatic, probably because of the presence of maternal antibodies (Haffejee 1995).

The highest prevalence of rotavirus gastroenteritis occurs in children between 6 and 24 mon of age, with the next highest prevalence in infants 1–6 mon of age, with neonates experiencing a low rate of rotavirus gastroenteritis (Kapikian 1996; Bartlett et al. 1988). Nevertheless, outbreaks of neonatal gastroenteritis caused by rotaviruses have been documented (Steele and Sears 1996).

During 1993–1995, 13.5% of hospitalizations among U.S. children aged 1 mon through 4 yr were associated with diarrhea (162,478/year). Rotavirus was the most common pathogen identified (16.5%), with an average of 26,798 cases per year (Parashar et al. 1998) (Fig. 1). Most hospitalizations caused by rotavirus occur in the 12- to 23-mon-old age group (Ferson 1996) (Fig. 2).

In a Connecticut study from 1987 through 1996, diarrhea-associated hospitalizations peaked from February through April, especially in children 4 mon to 3 yr of age. The apparent lower incidence peak was not observed among infants 1–3 mon of age (Parashar et al. 1999). This difference has been associated with the presence of protective maternal antibodies in children born from October through December (Newman et al. 1999).

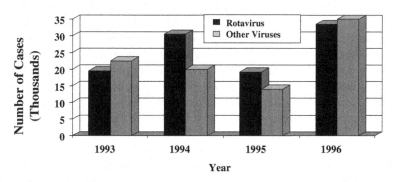

Fig. 1. Hospital admissions for enteric viruses in the U.S. (1993–1996) for those less than 15 years of age. Data from U.S. Department of Health and Human Services Vital and Health Statistics (National Hospital Discharge Survey) (Parashar et al. 1998).

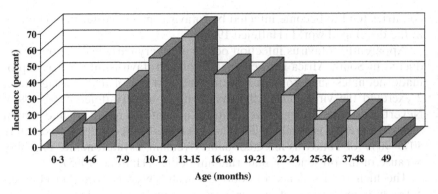

Fig. 2. Hospitalization of rotavirus gastroenteritis in the U.S. Data from Brandt et al. (1979).

Table 2. Rotavirus Prevalence in Selected Countries.

Country	Prevalence (%)	Reference
Australia	39.6	Ferson et al. 1996
England and Wales	43.0	Ryan et al. 1996
United States	39.0	Brandt et al. 1979
Hong Kong	28.5	Tam et al. 1986
Philippines	33.9	San Pedro et al. 1991
Mexico	49.0	Velazquez et al. 1993
Brazil	25.0	Stewien et al. 1991

In the U.S. from 1993 to 1996, rotavirus was identified as the cause of 10.4% of gastroenteritis-associated hospital admissions, increasing from 8.6% in 1993 to 14.7% in 1996. The annual incidence of rotavirus-associated hospitalizations was 4.4 per 10,000. The unadjusted median cost of a diarrhea-associated hospitalization during 1993–1996 was estimated to be $2,428 (Parashar et al. 1999). Each year in the U.S., rotavirus gastroenteritis results in $264 million in direct medical cost and more than $1 billion in total cost to society (CDC 1999c). The prevalence of rotavirus infection in children around the world varies from country to country, but most studies report a prevalence of 25%–50% (Table 2).

Enteric Adenoviruses. Enteric adenoviruses (Ead) are double-stranded DNA icosahedric viruses approximately 70 nm in diameter (Joki-Korpela and Hyypia 1998). At least 49 human adenoviruses have been identified (Calisher and Fauquet 1992). Adenoviruses may cause acute respiratory disease, pneumonia, epidemic conjunctivitis (Straus 1984), and acute gastroenteritis in children (Uhnoo et al. 1986). Documented waterborne outbreaks of conjunctivitis, by adenovirus type 3 (Martone et al. 1980; McMillan et al. 1992) and type 4 (D'Angelo et al. 1979), have been reported. Two outbreaks of gastroenteritis associated with drinking water have been reported (Kukkula et al. 1997; Divizia et al. 2004).

Ead 40 and 41 have been recognized as important etiological agents of gastroenteritis in children throughout the world (Uhnoo et al. 1986; Brandt et al. 1985; Leite et al. 1985; Albert 1986; Cruz et al. 1990; San Pedro and Waltz 1991), second in importance only to rotavirus. Where extensive studies have been conducted, it appears that Eads may be more important than rotavirus as a cause of diarrhea in developing countries (Herrmann and Blacklow 1995; Cruz et al. 1990).

Ead types 40 and 41 have been associated with outbreaks of gastroenteritis in day-care centers for children in the U.S. (Van et al. 1992). A large outbreak of keratoconjunctivitis within two day-care centers in Australia was caused by Ead type 8 (McMinn et al. 1991). An investigation in Guatemala (Cruz et al. 1990), showed that Ead40 (Reina et al. 1994; Parashar et al. 1999) and Ead41 were associated with diarrheal episodes in ambulatory children three times more often than rotaviruses. The incidence of adenovirus gastroenteritis in the world has ranged from 1.5% to 12.0%. Other types of adenoviruses have also been isolated from feces, but only Eads 40 and 41 have been consistently associated with gastroenteritis (Herrmann and Blacklow 1995).

In a 13-yr survey conducted in Australia in hospitalized children (0–14 yr of age), the Eads 40 and 41 were identified as the second most common cause of acute viral gastroenteritis, with an overall incidence of 6%; however, Ead infection was more common among infants (9.4%) (Barnes et al. 1998). In the Philippines, Eads have been associated with 5.4% of infant gastroenteritis, mainly during the rainy season (San Pedro and Waltz 1991).

Intussusception has been associated in some patients with adenovirus (types 1, 2, 5, and 6) infection. Mesenteric adenitis has been suggested as the probable cause; however, hyperirritability of the small intestine as a result of adenovirus infection has also been proposed as a possible cause of intussusception (Horwitz 1996). Eads may cause serious life-threatening illness in the immunocompromised (Gerba et al. 1996). For example, in cancer patients who are immunosuppressed the fatality rate for adenovirus infection is 53% (Hierholzer 1992). Similar fatality rates have been noted in bone marrow transplant patients.

There have been only two suspected drinking water outbreaks where an adenovirus may have been involved (Kukkula et al. 1997; Divizia et al. 2004). The lack of epidemiological evidence, the small number of studies, and limitations of methods of detection make this difficult to demonstrate. However, waterborne outbreaks of conjunctivitis and nose and throat infection by adenovirus types 3 and 4 are well documented (Martone et al. 1980; D'Angelo et al. 1979; McMillan et al. 1992). The Eads, in contrast to other adenoviruses, are not shed in respiratory secretions (Petric et al. 1982;

Blacklow and Greenberg 1991); thus, their transmission must be limited to the oral–fecal routes as with important waterborne pathogens (Beneson 1990).

Williams and Hurst (1988) reported that the number of indigenous adenoviruses detected in primary sewage sludge was 10 times greater than that of the enteroviruses. In addition, a greater number of adenoviruses than enteroviruses has been consistently found in raw sewage around the world (Irving and Smith 1981; Hurst et al. 1988; Krikelis et al. 1985a,b; Girones et al. 1993; Puig et al. 1994). Adenoviruses may survive longer in water than other enteric viruses (Enriquez et al. 1995).

The enteric nature of the adenoviruses 40 and 41, their presence only in the gastroenteric tract, and their extensive distribution suggest that water may play a role in the transmission of these agents. Furthermore, adenovirus type 31 has been increasingly detected during the last few years as an important cause of infant gastroenteritis (Thorner et al. 1993). Results of a comparative study of cytopathogenicity using immunofluorescence and *in situ* DNA hybridization as methods for the detection of adenoviruses from water, suggested that 80% of infectious adenoviruses in raw sewage may be Eads (Hurst et al. 1988).

As with other viral gastroenteritis, treatment of Ead diarrhea is directed at prevention of severe dehydration and electrolyte imbalance. Depending on the severity of dehydration, oral or intravenous rehydration may be needed (Hermann and Blacklow 1995).

Astrovirus. Astroviruses were first observed by electron micrographs of diarrheal stools by Madeley and Cosgrove (1975). These are icosahedral viruses with a starlike appearance and a diameter of approximately 28 nm. The human astrovirus group includes seven serotypes (Willcocks et al. 1994). Astrovirus type 1 seems to be the most prevalent strain in children. Type 4 has been associated with severe gastroenteritis in young adults. Astrovirus-like particles have been found in feces of a number of animals suffering from a mild self-limiting diarrheal infection, but no antigenic cross-reactivity has been found between these agents and human astroviruses.

Astrovirus infections occur throughout the year, with a peak during the winter/spring seasons in temperate zones in warm climates, but the highest incidence of astrovirus infection has been registered in May. Astroviruses cause a mild gastroenteritis after an incubation period of 3–4 d. Overt disease is common in 1- to 3-yr-old children. However, adults and young children are also affected. Astrovirus infection has been observed also in immunocompromised individuals and the elderly. Astroviruses are transmitted by the fecal–oral route. Outbreaks of astrovirus infection have been associated with oysters and drinking water (Kurtz and Lee 1987). An epidemiologic study in Guatemala showed that the astrovirus infection rate was 38% among children less than 3 yr of age, followed by Eads (22.3%), and rotaviruses (10.3%). Although astroviruses are common in developing countries, they have been identified also as a significant cause of infantile gastroenteritis in developed countries. A 10-yr study of astrovirus prevalence in Japan reported that 6%–10% of viral gastroenteritis cases in that country are caused by astroviruses. In England, these agents have been found in 7% of diarrheal samples examined by electron microscopy (Cubitt 1987). Astroviruses have also been associated with outbreaks in day-care centers for children (Mitchell et al. 1999).

A study in Glasgow reported that 80% of infants shedding astroviruses were suffering from gastroenteritis whereas 12% remained asymptomatic (Caul 1996b). Seroprevalence studies have indicated that by the age of 4 yr, 64% of all children have antibodies to astrovirus, with this number increasing to 87% in 5- to 10-yr-old children (Caul 1996b).

Caliciviruses. Caliciviruses are transmitted by the fecal–oral route. These viruses are important etiological agents of acute epidemic gastroenteritis that affect adults, school-age children, and family contacts (Kapikian 1996; Roper et al. 1990). Until recently, they were thought to only rarely infect children; however, recent studies (Koopmans et al. 2000; Inouye et al. 2000; Pang et al. 1992) suggest that they are a common cause of diarrhea in children <5 yr of age. In one study, the prevalence of diarrhea in children <5 yr exceeded that of rotavirus (Koopmans et al. 2000). It is now believed that the majority of previously undiagnosed cases of nonbacterial gastroenteritis is associated with caliciviruses (Glass et al. 2000). In a large study of outbreaks of infectious intestinal disease involving 40,000 cases in England and Wales, SRSVs (caliciviruses) were the most commonly identified agents (43% of cases), being more numerous than the cases of gastroenterititis caused by *Salmonella* and *Campylobacter* (Evans et al. 1998).

The family Caliciviridae is currently divided into four genera (Green et al. 2000). *Vesivirus* and *Lagovirus* contain only animal caliciviruses, which do not infect man. The human caliciviruses are divided into two genera, noroviruses and sapoviruses. The noroviruses are divided into two subgroups based on genotyping (Table 3).

Immunity to caliciviruses is poorly understood. Infectivity studies with volunteers have shown that immunity correlates inversely with serum or intestinal antibodies (Kapikian and Chanock 1996). Individual resistance to norovirus gastroenteritis is more important than acquired immunity (Blacklow et al. 1987). It has been suggested that genetically determined factors are the primary determinants of resistance to norovirus infection, perhaps at the level of cellular receptor sites (Blacklow et al. 1987). In developing countries, antibodies to norovirus are acquired early in life, and peak incidence of illness may also occur among younger age groups than that in developed nations (Roper et al. 1990). It has been proposed that noroviruses may circulate in low numbers in a population until an infected individual contaminates a common source of food or water, resulting in

Table 3. Human Calicivirus Genera and Representative Strains.

Genus and genogroup	Virus common name
Noroviruses	
Genogroup I	Norwalk virus
	Southhampton virus
	Desert Shield virus
	Cruise ship virus
Genogroup II	Snow Mountain agent
	Hawaii virus
	Toronto virus
	Lordsdale virus
	Grimsby virus
	Gwyneld virus
	White River virus
Sapporoviruses	Sapporo virus
	Manchester virus
	Parkville virus
	London virus
	Houston virus

explosive outbreaks (Roper et al. 1990). Noroviruses and related viruses usually produce a mild and brief illness, lasting 1–2 d, characterized by nausea and abdominal cramps, followed commonly by vomiting in children and diarrhea in adults (Gouvea et al. 1994). The involvement of norovirus as a pathogen of adults was further suggested by Numata et al. (1994), who reported a very low prevalence of antibodies to recombinant norovirus capsid protein in children <7 yr of age, but an increasing prevalence in individuals from 12 to 50 yr of age in Japan. Recent studies suggest that they may be the most common cause of foodborne outbreaks (Inouye et al. 2000; Deneen et al. 2000).

Oral rehydration is generally sufficient to treat Norwalk virus and calicivirus gastroenteritis. In rare cases, intraveneous administration of liquids and electrolytes may be necessary (Estes and Hardy 1995).

B. Hepatitis Viruses

Hepatitis A Virus (HAV). HAV is a picornavirus that is morphologically indistinguishable from other members of the same family (Hollinger and Ticehurst 1996). Relatively resistant to heat, it is partially inactivated after 12 hr at 60°C. Infectivity at room temperature is maintained for 1 mon after drying, and the virus can survive for days to months in different types of water (Hollinger and Ticehurst 1996). Each year, approximately 140,000

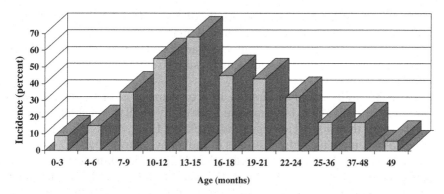

Fig. 3. Incidence of hepatitis A by age in the U.S. (1997). Data from Brandt et al. (1979).

Table 4. Sources of Hepatitis A Virus (HAV).

Source	Percent
Personal contact	25–30
Day-care centers	10–15
Contaminated food and water	3–8
Travel to endemic areas	9
Unknown	50

persons in the U.S. are infected with HAV with an approximate annual cost of $200 million (Fishman et al. 1996). A more recent study placed the costs for adolescents (>15 yr of age) and adults between $332 and $580 million a year (Berge et al. 2000). The highest rates of disease are among persons 5–14 yr old (CDC 1999b) (Fig. 3).

The incubation period for HAV is approximately 28 d. The infected individual sheds HAV actively during the initial stage of the infection period, 1–2 wk before overt disease (Fishman et al. 1996). Propagation of HAV is mainly fecal–oral because hepatitis A is often associated with unsanitary and crowded conditions; however, bloodborne transmission may be possible during the viremic phase that occurs in the initial stage of the disease (Fishman et al. 1996). Personal contact accounts for 25%–30% of HAV cases in the U.S., followed by day-care centers (10%–15%), contaminated food or water (3%–8%), and travel to endemic areas (9%) (Fishman et al. 1996) (Table 4). In almost half the cases, no source of exposure can be documented.

Hepatitis A is usually a mild illness, which almost always results in complete recovery. Severity and disease manifestation are age related. An estimated 80%–95% of infected children younger than 5 yr of age do not develop overt disease, whereas clinical manifestations are observed in approximately 75%–90% of infected adults (Harrison 1999). Neither chronic hepatitis nor a carrier state results from HAV (Fishman et al. 1996).

The mortality rate in children of 14 yr or younger is 0.1%; this rate rises to 0.3% in individuals between the ages of 15 to 39 yr, reaching 2.1% in those older than 40 yr (Hollinger and Ticehurst 1996). Although hepatitis A is usually benign in children, studies in India have shown that 10% of cases of acute liver failure are associated with HAV infection. Furthermore, HAV coinfection with hepatitis E virus (HEV) accounts for 22.5% of acute liver failure in children, with the 5- to 10-yr-old group being the most affected. The mortality rate observed in these patients was 63.6% (Arora et al. 1996).

In developing countries, the incidence of symptomatic hepatitis A in adults is relatively low because of exposure to the virus in childhood. Most individuals 18 yr and older possess an immunity that provides lifelong protection against reinfection (Hollinger and Ticehurst 1996). In the U.S., the percentage of adults with immunity increases with age (10% for those 18–19 yr of age to 65% for those >50) (Margolis et al. 1997).

In areas with poor sanitation, nearly all children up to 9 yr of age have been infected by HAV. In these areas, outbreaks rarely occur, and clinical disease related to HAV infection is uncommon. Under better sanitation, HAV infection shifts to older individuals, and the incidence of overt disease increases.

Hepatitis A is one of the few viral diseases that has been clearly proved to be transmitted by the water route. In a retrospective study of waterborne disease outbreaks, occurring in the U.S. from 1946 to 1980, Lippy and Waltrip (1984) reported that viral hepatitis was frequent, with 68 outbreaks and 2,262 cases. Many outbreaks have been traced to the consumption of raw shellfish (Rao and Melnick 1986).

Treatment of hepatitis A is usually supportive, although passive immunization with pooled human immunoglobulin (Ig) containing high titers of anti-HAV has been the only means to provide preexposure or postexposure immunoprophylaxis against hepatitis A. Although a hepatitis A vaccine was first licenced in 1995, hepatitis A continues as one of the most frequently reported vaccine-preventable diseases in the U.S. (CDC 1999b). Initially, immunization against HAV was mainly of children living in communities with the highest rates of HAV infection (CDC 1996). To reduce HAV incidence, widespread vaccination of appropriate susceptible populations needs to be implemented. The Advisory Committee on Immunization Practices (ACIP) recently recommended the routine vaccination of children from communities with rates that are twice the 1987–1997 national average (10–19 cases per 100,000 population) and the consideration of routine vaccination of children from communities with rates higher than the 1987–1997 national average (CDC 1999b).

Two inactivated HAV vaccines have been licensed in the U.S. (CDC 1999b). HAV exists as a single serotype and exposure to it induces a life-long immunity. In HAV-endemic areas, children of 10–15 yr of age are likely to be HAV seropositive. Due to the high HAV antibody prevalence in adults in the U.S., pooled sera contain titers of anti-HAV immunoglobulins that are protective (Fishman et al. 1996).

Hepatitis E Virus (HEV). The hepatitis E virus (HEV), formerly known as enterically transmitted non-A, non-B hepatitis virus, is the leading etiology of acute viral hepatitis in developing countries (Bradley et al. 1992; White and Fenner 1994). HEV accounts for more than 50% of acute hepatitis cases in Asia and Africa and is the most common cause of hepatitis in children (Fishman et al. 1996). In the U.S. and in Europe, most hepatitis E cases have been confirmed only in people returning from endemic areas (Fishman et al. 1996).

HEV was originally placed in the Calicivirdae family but has now been removed (Berke and Matson 2000). Its taxonomic position is now uncertain. Isolates of this agent can be broadly grouped into two serotypes, Mexico-HEV and Burma-HEV, the latter being more prevalent. In experimental trials, primates infected with the Burma isolate were protected from infection by the Mexico-HEV. This finding indicates that these two serotypes share neutralizing antigens (Bradley et al. 1992). Balayan (1993) has suggested that the higher prevalence of HEV in adults may result from a silent infection early in life, with subsequent waning of immunity after 10–20 yr, becoming again susceptible to infection by HEV at a later age. An important epidemiological feature of HEV infection is the frequent occurrence of outbreaks associated with consumption of sewage-polluted water (Balayan 1993).

In contrast to hepatitis A, hepatitis E occurs in young and middle-aged adults. In addition, most cases originate from a primary source, with infrequent cases among secondary contacts, compared to hepatits A in which close contact is a common risk factor (Harrison 1999). Transmission of HEV is fecal–oral, often through contaminated water and food (Tan et al. 2003). Outbreaks often occur during the rainy season, monsoons, or flooding, with sporadic cases resulting from person-to-person transmission (Fishman et al. 1996). Current evidence suggests that HEV is zoonotic and can be transmitted by undercooked deer, pork, and boar meat (Tei et al. 2004).

The incubation of HEV is 2–8 wk with an average of 5–6 wk, which is slightly longer than HAV (White and Fenner 1994). The most evident pathologic feature of hepatitis E, in contrast to hepatitis A, is cholestasis. Hepatitis E is a self-limiting, acute disease. Similar to hepatitis A, its severity increases with age; this may explain why the disease is mostly reported among young adults, whereas children are usually unaffected. However, simultaneous infection of young children with HEV and HAV has resulted in severe disease, sometimes with acute liver failure (Arora et al. 1996). The disease is most often seen in older children to middle-aged adults (15–40 yr old). The disease is often mild and resolves in 2 wk, leaving no sequela; however, the fatality rate may be high (2%–3%) (Haas et al. 1999). Furthermore, pregnant women appear to be exceptionally susceptible to severe disease, and the fatality rate may reach 17%–33% (Haas et al. 1999). There is no evidence of immunity against HEV in the population that has been exposed to this virus (Margolis et al. 1997).

HEV has been isolated from pigs in the U.S., and this isolate has been shown to be infectious to primates. In addition, the pig isolate is highly homologous, at the nucleotide level, to HEV strains isolated from humans (Harrison 1999).

Unlike hepatitis A, specific antibodies do not prevent or mitigate the clinical manifestation of hepatitis E; therefore, only the implementation of appropriate sanitary practices and the consumption of uncontaminated drinking water and food may diminish the risk of HEV infection (Fishman et al. 1996). Treatment is only supportive of the clinical symptoms.

C. Enteroviruses

Enteroviruses (polioviruses, coxsackieviruses, echoviruses, and enteroviruses) are among the most common and significant causes of infectious illness in infants and children. Non-polio enteroviruses are estimated to cause 10–15 million symptomatic infections in the U.S. annually (Zaoutis and Klein 1998). They are associated with a broad spectrum of clinical syndromes, including aseptic meningitis, herpangina, hand-foot-mouth disease, conjunctivitis, pleurodynia, myocarditis, poliomyelitis, various exanthems (rashes), and nonspecific febrile illness. While all age groups can become affected, the most serious outcomes are in the newborns, young children, and adults. Poliomyelitis, once a common crippling disease, largely in older children in the U.S., has largely been eliminated around the world due to the development of poliovirus vaccines in the 1950s. Because enteroviruses were the first human viruses grown in cell culture, a great deal has been learned about their epidemiology. Infections are most common in childhood. Isolation of echovirus and coxsackievirus from stools of children may be as high as 8%–10% during the summer months (Fox and Hall 1980). The fecal–oral route is believed to be the main route of transmission, although respiratory transmission may also be significant for some types (Morens et al. 1991). It is believed that almost all enteroviruses (except possibly enterovirus type 70, which causes eye infections) can be transmitted by the fecal–oral route.

Incubation periods vary greatly with the type of virus and may be as short as 12 hr for coxsackievirus type A24 (eye infections) to as long as 35 d for poliovirus. The presentation of symptomatic illness is also highly type- and strain dependent. Some enterovirus infections may pass through a community with no illness observed (Fox and Hall 1980). Echoviruses usually cause milder illnesses than those of coxsackievirus. The overall case fatality ratio in recognized cases of enterovirus illness has been reported to range from 0.01% to 0.94% (Assad and Borecka 1977). The incidence of serious neonatal coxsackievirus infections is about 1 in 2,000 live births, 10% of which are usually fatal (Kaplan et al. 1983). Such infections are usually acquired by the mother and transmitted to the child after birth. Because of the large variety of symptomatic to asymptomatic cases among enterovirus types, long incubation periods, a wide variety of symptoms, and costly isolation methods, it has been difficult to document common-source outbreaks.

Newer technologies, such as polymerase chain reaction (PCR), are rapid and sensitive testing methods for diagnosis of enteroviral infections, which may expand the list of diseases attributable to this group of pathogens. Although treatment of enteroviral infections remains unsatisfactory, immunization against poliovirus has been remarkably successful.

Properties of the Enteroviruses. The enteroviruses are a subgroup of single-stranded RNA, nonenveloped viruses belonging to the Picornaviridae family (pico = small, RNA = ribonucleic acid). They include the polioviruses, coxsackieviruses, echoviruses (echo = enteric cytopathogenic human orphan), and unclassified enteroviruses. Early classification of enteroviruses involved groupings based on cytopathological effect in tissue culture. Newly discovered enteroviruses are now simply assigned enterovirus type numbers. The enteroviruses currently recognized to infect humans are outlined in Table 5.

The virion consists of an icosahedron-shaped protein capsid and an RNA core. Although the capsid proteins determine antigenicity, there are no significant antigens common to all members of this group of viruses. The virus can withstand the acidic pH of the human gastrointestinal tract and can survive at room temperature for several days. These features enable the fecal–oral mode of transmission. Most enteroviruses can grow in primate (human or nonhuman) cell cultures, exhibiting cytopathic effects. Enteroviruses are commonly referred to as "summer viruses" because resulting infections occur primarily during the warmer, summer months

Table 5. Human Enteroviruses.

Group	Serotypes
Poliovirus	1–3
Coxsackievirus group A	1–22, 24
Coxsackievirus group B	1–6
Echovirus	1–9, 11–27, 29–33
Enterovirus	68–91

(May through October) in temperate northern hemisphere climates such as in the U.S. In tropical climates, enteroviral infection is seen all year without seasonal variation. Humans are the only known natural hosts for enteroviruses.

The fecal–oral route is the most common mode of transmission, but oral–oral and respiratory spread are also possible. Risk factors for infection include poor sanitation, crowded living conditions, and low socioeconomic class. Children <5 yr of age are the most susceptible to infection, partly because of a lack of prior immunity and the poor hygienic habits associated with this age group.

Enterovirus Illness. The incubation period for most enteroviral infections ranges from 3 to 10 d. The virus enters the host via the oral and/or respiratory tract, then invades and replicates in the upper respiratory tract and small intestine, with a predilection for lymphoid tissue in these regions (Peyer patches, mesenteric nodes, tonsils, and cervical nodes). Virus then enters the bloodstream, resulting in a minor viremia and dissemination to a variety of target organs, including the central nervous system (CNS), heart, liver, pancreas, adrenal glands, skin, and mucous membranes.

The infections cause a wide spectrum of disease that can involve almost any organ system (Table 6). Disease severity can range from life threatening with significant morbidity to mild or subclinical. It is believed that approximately 50% of non-polio enterovirus infections are asymptomatic. The more common syndromes include nonspecific febrile illness, aseptic meningitis, herpangina, hand-foot-mouth syndrome, and exanthems. The clinical manifestations of infection in the neonate can be distinct and are discussed separately. Currently, there are no vaccines (except poliovirus) or treatment of enterovirus infections, except those supportive of clinical symptoms.

Paralysis.   Before the advent of vaccination, poliovirus was a major cause of permanent paralysis in the U.S. Vaccination has largely eliminated poliomyelitis in the U.S., and no indigenous wild viruses have been detected in the U.S. since 1979 (Morens et al. 1991). Most infections are asymptomatic (90%–95%), with only 0.1%–2% resulting in paralytic poliomyelitis. Most poliovirus infections occur in children <4–5 yr of age, but the older the age of infection, the greater the severity of the outcome. Mortality in children averages 2.5% for symptomatic infections and 30% in adults (Morens et al. 1991).

Non-polio enteroviruses have been associated with paralysis, but this is uncommon compared to poliovirus (Gauntt et al. 1985; Grist and Bell 1984). Coxsackievirus A7 has been associated with outbreaks of paralytic disease (Grist and Bell 1970), and outbreaks of enterovirus 71 have been involved in several outbreaks of CNS involvement, with fatal cases mostly in children (Melnick 1984).

Table 6. Common Clinical Syndromes Associated with Enterovirus Infections in Children.

Syndrome	Predominant Virus	Clinical Features
Nonspecific febrile illness	All types	Febrile illness (fever), with nonspecific upper respiratory and gastrointestinal tract symptoms
Aseptic meningitis	Echovirus, group B coxsackieviruses, and polioviruses	Fever, meningeal signs with mild cerebrospinal fluid (CSF) pleocytosis, usually normal CSF glucose and protein, and absence of bacteria
Herpangina	Group A coxsackieviruses	Fever, painful oral vesicles on tonsils and posterior pharynx
Hand-foot-mouth disease	Coxsackievirus A16	Fever, vesicles on buccal mucosa and tongue and on interdigital surfaces of hands and feet
Nonspecific exanthem	Echoviruses	Variable rash (usually rubelliform but may be petechial or vesicular), with or without fever
Pleurodynia	Coxsackievirus B3, B5	Uncommon, epidemic, fever, and severe muscle pain of chest and abdomen
Myocarditis	Group B coxsackieviruses	Uncommon, myocarditis/ pericarditis, which can present with heart failure or dysrhythmia
Acute hemorrhagic conjunctivitis	Enterovirus 70	Epidemic cause of conjunctivitis with lid swelling, subconjunctival hemorrhage, and eye pain without systemic symptoms
Paralytic disease	Poliovirus, enterovirus 71, echoviruses, and coxsackieviruses	Paralysis

Perinatal and Neonatal Infections. Neonates represent a population at great risk from severe enteroviral disease. Adverse effects also occur from enteroviral infection during pregnancy with adverse effects to the fetus. In the prevaccine era, paralytic poliomyelitis occurred during pregnancy in apparent excess of age-adjusted expected rates, suggesting predisposition among pregnant women (Abzug et al. 1995). Infections of pregnant women by the non-poliovirus enteroviruses occur frequently. In a seroepidemiologic study, Brown and Karunas (1971) found a 42% rate of infection during pregnancy in a population evaluated prospectively over a 10-yr period. In a review of coxsackie B infections, Modlin and Rotbart (1997) suggested that greater viral replication and prolonged maternal enterovirus excretion occurring in late pregnancy may well enhance the risk of infection of the newborn infant in the perinatal period.

Neonatal non-polio enteroviral infections are common. Estimated attack rates indicate that disease in newborns and young infants is comparable or exceeds symptomatic neonatal infections caused by herpes simplex virus and cytomegalovirus (Jenista et al. 1984; Kaplan et al. 1983; Modlin 1986). Enteroviruses were responsible for the majority (65%) of >3 mon-old infant admissions to one hospital in one community for suspected sepsis (Dagan et al. 1989; Kaplan et al. 1983; Modlin 1986). In another study, enteroviruses were the most frequently identified pathogen between days 8 and 29 of life, accounting for at least one-third of all cases of neonatal meningitis (Dagan et al. 1989; Kaplan et al. 1983; Modlin 1986; Shattuck and Chonmaitree 1992).

Age is one of the most important determinants of outcome of enterovirus infections. Different age groups have different susceptibilities to infection, different clinical manifestations and degrees of severity, and different prognoses following enteroviral infection. Young children have higher attack rates. In one study, echovirus 9 disease attack rates in children were found to be 50%–70% compared with 17%–33% in adults (Lerner et al. 1963). Age-specific attack rates of echovirus 30 per 1,000 persons in an outbreak in the United Kingdom ranged from 19.7 (children age 0–9 yr) to 7.11, 4.82, 4.73, 1.5, and 0 for the succeeding 10-yr age cohorts, respectively (Irvine et al. 1967).

Severity of illness may also be age dependent. With poliovirus infection, adults are more likely to be severely affected, tending to acquire paralytic poliomyelitis rather than nonparalytic poliomyelitis (i.e., aseptic meningitis) or asymptomatic infections. On the other hand, coxsackie B virus infection is clearly more severe in newborns than in older children and adults, often causing myocarditis, encephalitis, hepatitis, and death (Eichenwald et al. 1967; Gear and Measroch 1973; Woodruff 1980). Coxsackievirus and echovirus encephalitis and aseptic meningitis are most frequent among those 5–14 yr old (Ball 1975; Forbes 1963; Karzon et al. 1961), while myocarditis is most common in adults and neonates. In another study (Dery et al. 1974), the mean age among patients with coxsackievirus B meningitis was 7.7 yr, pericarditis was 9.9 yr, and gastroenteritis was 1.3 yr.

Herpangina.   Herpangina is characterized by a painful vesicular eruption of the oral mucosa associated with fever, sore throat, and pain on swallowing. It is seen most commonly in children ages 3–10 yr (Morens et al. 1991). Group A coxsackieviruses are the most common etiological agents, but group B coxsackieviruses and echoviruses also have been isolated from patients. Fever, usually mild, develops suddenly, but higher temperatures up to 41°C (105.8°F) can be seen, particularly in younger patients. Nonspecific early symptoms may include headache, vomiting, and myalgia. Sore throat and pain with swallowing are the most prominent symptoms and precede the characteristic exanthem (eruption of mucous membranes) by approximately 1 d. Herpangina is self-limiting, and symptoms resolve within 1 wk. Young children are at risk for dehydration because of refusal to eat or drink.

Aseptic Meningitis. Non-polio enteroviruses are the leading causes of aseptic meningitis, accounting for 70%–90% of all cases from which an etiological agent is identified (McGee and Barmger 1990; Zaoutis and Klein 1998). The most common enterovirus types associated with aseptic meningitis are coxsackievirus B5 and echovirus 4, 6, 9, and 11. These have occurred in epidemic outbreaks as well as sporadic cases, being most common in the 5- to 15-yr age group (Morens et al. 1991). Outbreaks have been associated with child-care centers (Helfand et al. 1994; Mohle-Boetani et al. 1999).

The initial presentation of enteroviral meningitis is similar to that of nonspecific febrile illness. Commonly, a biphasic pattern of symptoms is seen, with signs of CNS involvement in addition to recurrence of fever. Evidence for meningeal irritation commonly includes headache and photophobia, with 50% of children >1–2 yr of age also developing a stiff neck.

The course of enteroviral meningitis usually is self-limiting and benign, but there has been an ongoing debate about the occurrence of long-term sequelae (Modlin and Rotbart 1997). Recent studies have shown that there are no long-term neurological, cognitive, or developmental abnormalities from this infection in older children (Rorabaugh et al. 1993). However, several investigations have documented that 10% of children <3 mon of age who have aseptic meningitis may suffer long-term sequelae, especially speech and language delay (Etter et al. 1991).

Nonspecific Febrile Illness. The most common clinical presentation of non-polio enterovirus infection is a nonspecific febrile illness (fever). Typically, fever develops suddenly, and temperatures range from 38.5°C to 40°C (101°F to 104°F) and last an average of 3 d. Occasionally, a biphasic pattern of symptoms can be seen, with an initial fever for 1 d, followed by 2–3 d of normal temperatures and recurrence of fever for an additional 2–4 d.

Studies have found that enteroviruses are the major cause of hospitalization for young infants (<2–3 mon of age) for suspected fever caused by septicemia during the summer and fall (Dagan 1996). A recent study of infants <90 d of age found that non-polio enteroviruses were the most common cause of fevers in infants requiring hospitalization (Byington et al. 1999). More than 25% of the infants were infected; the average stay was 3 d with average medical costs of $4,500.

Exanthems (Eruption of the Skin). Non-polio enteroviruses are the leading cause of exanthems in children during the summer and fall months. The most common serotype causing exanthem is echovirus 9. The classic enteroviral exanthem consists of a pink, macular, rubelliform rash. The rash may be the sole manifestation of infection or may be present in association with febrile illness or aseptic meningitis. Enteroviral exanthems are seen most commonly in children <5 yr of age and decrease in prevalence with age. The rash is self-limiting and disappears in 3–5 d (Zaoutis and Klein 1998). Most infections occur in infants and young children 3–10 yr of age (median, 4 yr) (Morens et al. 1991). An outbreak has been reported in a day-care center (Moreira et al. 1995).

The best known enteroviral exanthem is hand-foot-mouth (HFM) disease. It is commonly associated with coxsackievirus A16, but may also be caused by coxsackievirus A5 and several other enteroviruses including enterovirus 71 (Hagiwara et al. 1978). Children usually have a fever, with multiple discrete red macular lesions of about 4 mm appearing on the palms, soles, fingers, and toes. The infection is usually self-limiting, lasting 1–2 wk.

Complications associated with HFM disease caused by enterovirus type 71 include encephalitis, meningitis, hemorrhage, acute flaccid paralysis, and myocarditis. During a large outbreak of HFM disease in Taiwan in 1990 of enterovirus type 71, 129,000 cases were estimated, resulting in 405 hospitalizations and 78 deaths (Ho et al. 1999). Almost all the severe cases and deaths were in children <5 yr of age.

Respiratory Illness. Worldwide enteroviruses appear to account for 2%–15% of all viruses that cause upper and lower respiratory tract diseases (Chonmaitree and Mann 1995). The illness is most commonly associated with coxsackie A10, A21, A24, and B2 (Moren et al. 1991). Both children and adults are affected, with infections lasting only a few days. Pneumonia associated with enteroviral infection has been reported in both outbreaks and individuals (Chonmaitree and Mann 1995). Fatal pneumonia has been associated with coxsackievirus and echovirus infections in infants and children (Boyd et al. 1987; Cheeseman et al. 1977; Craver and Gohd 1990).

Acute Hemorrhagic Conjunctivitis (AHC). This explosive epidemic conjunctivitis, first described in 1969 in Africa and Asia, is now found worldwide. It is common in tropical and densely populated regions. The majority of outbreaks have been caused by enterovirus serotype 70, but recently coxsackievirus A24 has been isolated during outbreaks (Morens et al. 1991). AHC is characterized by sudden onset of severe eye pain, photophobia, and blurred vision. Subconjunctival hemorrhages, erythema, edema of the lids, and eye discharge are characteristic of infections. Recovery occurs within 7–10 d. Spread is by the eye-hand-fomite route, in contrast to the fecal–oral route seen with most enteroviral infections. Overall, it is more common in adults, but it also does affect school-age children. Some enterovirus outbreaks have been associated with poliomyelitis-like paralysis.

Diabetes. Insulin-dependent diabetes mellitus (IDDM) is the most common severe chronic childhood illness, affecting an estimated 123,000 children in the U.S. (Libman et al. 1993). More than 11,000 new cases are diagnosed annually. The disease is the leading cause of renal failure, blindness, and amputation and a major cause of cardiovascular disease and premature death in developed countries (Rewers and Atkinson 1995). IDDM occurs most frequently at the ages of 2, 4–6, and 10–14 yr, perhaps because of physiological increases in sex hormone levels and insulin resistance or because of alterations in the pattern of childhood infections. Season and latitude affect incidence, suggesting an infectious etiology (Rewers and Atkinson 1995). The infectious agents most commonly linked to IDDM have been the enteroviruses.

To date, epidemiological studies have failed to prove or disprove the association of enteroviruses with IDMM (Green et al. 2004), perhaps because the nature of the disease may involve both genetic factors of the host and environmental exposure, with clinical symptoms taking years to develop. Autoimmunity, potentially induced by a preceding enterovirus infection, could play a role in human IDDM (Rewers and Atkinson 1995). Recent studies continued to support some association with enterovirus infections and IDDM (Nairn et al. 1999; Pallansch 1997; Smith et al. 1998; Knip and Akerblom 1998, 1999; Hoyoty et al. 1998, Honeyman et al. 2000; Lonnrot et al. 2000).

Pleurodynia (Bornholm Disease). Pleurodynia or epidemic myalgia is characterized by an acute onset of severe muscular pain in the chest and abdomen accompanied by fever. Coxsackieviruses B3 and B5 are the major causes of epidemic disease; rare sporadic cases have been described with other non-polio enteroviruses.

The muscular pain is sharp and spasmodic, with episodes typically lasting 15–30 min, although they can last up to several hours. During spasms, patients may develop signs of respiratory distress or appear shocklike with diffuse sweating and pallor. Pain localized to the abdomen in young children may falsely suggest intussusception or appendicitis. The illness usually lasts 1–2 d, but frequent recurrences are possible several weeks after the initial episode. Associated signs and symptoms include anorexia, headache, nausea, and vomiting. In contrast to many other enteroviral syndromes, pleurodynia is more common in older children and adolescents.

Cases are recognized mostly in school-age children and adults, with the peak age being children 2–9 yr old (Morens et al. 1991). However, older boys have also been reported to develop orchitis or inflammation of the testes (Morens et al. 1991). It has not been established whether involvement of the ovaries occurs.

Myocarditis. Coxsackievirus B infections are increasingly being recognized as a cause of primary myocardial disease in adults as well as children

(Melnick 1997). In some studies, up to 39% of persons infected with coxsackievirus B5 developed cardiac abnormalities. Coxsackieviruses of group A and echoviruses have also been implicated, but to a lesser degree. The illness is common in neonates and adults. Older adults represent the vast majority of cases, with patients aged 40 and older composing 69% of the cases (Martino et al. 1995). Although the incidence is less in neonates, the outcome is potentially more severe, with mortality among infants reported to be 30%–50% (Modlin and Rotbart 1997). Symptoms usually begin within the 10th day of birth with fatigue, poor feeding, or mild respiratory distress.

Most children and adults recover; however, one or more recrudescences several weeks to more than a year later have been reported in approximately 20% of the cases after the initial illness (Modlin 1990). Persistent electrocardiographic abnormalities (10%–20%), cardiomegaly (5%–10%), and chronic congestive heart failure indicate that permanent heart damage occurs as a result of this illness.

Diseases Associated with Immunocompromised Children.   Enteroviruses are not prominent among the microorganisms that cause serious morbidity and mortality among the immunocompromised. In childhood, serious enterovirus infection does not appear to be particularly common in the T-cell immunodeficiency syndromes (Morens et al. 1991). However, enteroviral infections pose significant risk to children who have defects in B-lymphocyte function, the most common of which is X-linked agammaglobulinemia (Gewurz et al. 1985; McKinney et al. 1987; Hertal et al. 1989). Unlike other viruses that are combated by cellular immune mechanisms, enteroviruses are eliminated from the host by humoral immune mechanisms. An intact B-cell response is believed to be necessary to block viral entry into the CNS. Children who have agammaglobulinemia may develop chronic enteroviral infection, most commonly meningoencephalitis. Patients experience headache, lethargy, seizures, motor dysfunction, and altered sensorium. Symptoms may wax and wane for years, but there is an overall progressive deterioration in CNS function. Infections are fatal in most children who are immunodeficient. Echovirus 11 has been the most common cause of chronic infection, but cases caused by other echoviruses and coxsackieviruses have been reported (Morens et al. 1991).

Enterovirus infections in infants who have received organ transplants can result in serious complications (Chuang et al. 1993; Aquino et al. 1996). Serious life-threatening infections of both echovirus and coxsackievirus have been documented in infants receiving both bone marrow and liver transplants. Children cancer patients receiving chemotherapy may also suffer from severe illness when infected with a coxsackievirus (Geller and Condie 1995).

Other Illnesses.   Enteroviruses have been associated with a number of other illnesses that affect children (Kennedy et al. 1986), including juvenile

Table 7. Less Common Illnesses Associated with Enterovirus Infections in Children.

Syndrome	Reference
Rheumatoid arthritis	Blotzer and Myers 1978
	Heaton and Moller 1985
	Zaher et al. 1993
Pancreatitis	Kennedy et al. 1986
Hemorrhagic syndrome	el-Sageyer et al. 1998
Gastroenteritis	Birenbaum et al. 1997
Hepatitis	Jeffery et al. 1993
Mental Disorders	Hirayama et al. 1998
Alice in Wonderland syndrome	Wang et al. 1996
Schizophrenia	Rantakallio et al. 1997
Vertigo	Simonsen et al. 1996
Hydrancephaly (absence of cerebral hemisphere in the newborn)	Marlin et al. 1985

rheumatoid arthritis (Blotzer and Myers 1978) and gastroenteritis (Joki-Korpela and Hyypia 1998). Case reports have also linked enteroviruses to short-term mental impairment in children and other illnesses or symptoms in children (Table 7). Other studies have suggested relationships between enterovirus infections and sudden infant death syndrome (SIDS) (Rambaed et al. 1999), risk of schizophrenia from infections early in childhood (Rantakallio et al. 1997), amyotrophic lateral sclerosis (Lou Gehrig's disease) (Berger et al. 2000), vertigo (Simonsen et al. 1996), and chronic fatigue syndrome (Galbraith et al. 1997; Lane et al. 2003). These studies have been limited in scope or speculative.

III. Incidence of Enteric Virus Infection by Age
A. Rotavirus

Rotavirus is the major cause of childhood gastroenteritis, although all age groups are affected. The highest incidence of the disease is in the fall and winter in the U.S. In one study rotavirus was detected in 29% of the stools of children <2 yr of age, with 48% of the cases being asymptomatic (Champsaur et al. 1984a,b). In another study the incidence of rotavirus gastroenteritis was found to be 40% in the 1–2 yr age group, 12% in the 2–3 yr age group, and 5% in adults (Rodriguez et al. 1987). Crowley et al. (1997) found that almost 65% of the diagnosed cases in England and Wales occurred in the 6-mon to 1-yr age group (Table 8).

Table 8. Distribution (Percent) by Age of Gastroenteritis Infections in Small Children in England and Wales, 1990–1994.

Virus	<1 mon	1–2 mon	3–5 mon	6–11 mon	1 yr	2 yr	3 yr	4 yr	Total number of cases
Rotavirus	1.6	4.9	10.8	29.3	35.3	12.2	4.2	1.7	6,591
Adenovirus	1.6	8.3	15.7	27.2	27.5	11.5	5.4	2.8	10,362
SRSV	1.5	5.0	12.7	27.5	29.8	12.9	6.9	3.6	1,756
Astrovirus	0.5	4.6	12.5	25.0	31.1	16.5	6.3	3.5	1,760

Source: Crowley et al. (1997).

B. Adenovirus

The incidence of adenovirus gastroenteritis in the world has ranged from 1.5% to 12.0% (Herrmann and Blacklow 1995). In England, a survey of stool samples from patients suffering from viral gastroenteritis showed that enteric adenoviruses were present in 14% of the examined specimens. Recent studies in England and Wales (Caul 1996b) have indicated that enteric adenoviruses account for 8.2% of all viral gastroenteritis. Although most reports indicate that the enteric adenoviruses are only second in importance to rotavirus as a cause of viral gastroenteritis, an epidemiological study in Guatemala showed that the adenoviruses 40 and 41 were associated with diarrheal episodes in ambulatory children three times more often than rotaviruses (Cruz et al. 1990).

C. Caliciviruses

Recent research suggests that calicivirus diarrhea may be common among infants and young children (Koopmans et al. 2000; Inouye et al. 2000; Pang et al. 1992). Koopmans et al. (2000) reported that it was a more common cause of gastroenteritis in children <5 yr than rotavirus. Norwalk virus antibodies are acquired gradually, beginning slowly in children, and increasing in adulthood. By age 50, approximately half the population has developed antibodies to Norwalk virus (Estes and Hardy 1995).

D. Hepatitis A Virus (HAV)

Infection of HAV in children is usually asymptomatic; however, the risk of symptomatic cases increases to 75% in adults in whom the most severe cases are seen. The incidence of reported hepatitis A in the U.S. is 9.7 cases per 100,000 (CDC 1996). However, the actual incidence may be much higher because many persons do not seek treatment and because physicians are believed to report fewer than 15% of hospital diagnosed cases (Hollinger and Ticehurst 1996). Thus, the true incidence is at least 6.6 times

Table 9. Percentage of Stool Samples Positive for Enteric Viruses.

Virus	<1 yr	1–4 yr	5–14 yr	15–74 yr	>74 yr
Rotavirus group A	10.3	8.8	4.5	2.0	0.0
Adenovirus types 40 and 41	7.1	6.1	1.8	0.0	0.0
Astrovirus	7.1	4.0	0.9	1.2	0.0
SRSV[a] and Calicivirus	17.8	13.5	9.9	5.2	4.5

[a] Small round structured viruses (Tompkins et al. 1999).

that reported (i.e., 0.1%). Also, this incidence does not take into consideration the number of asymptomatic cases, which occur commonly among young children. In an endemic area of Italy, De Filippis et al. (1987) found that 8.2% of stool samples from healthy individuals contained HAV, with the highest prevalence found in children. The greatest incidence is among children 5–14 yr of age. It has been estimated that 95,000–180,000 infections occur yearly in children <10 yr of age in the U.S. (Armstrong and Bell 2002).

E. Hepatitis E Virus (HEV)

In a survey in Mexico, where HEV is endemic, it was found that 10.5% of 3,549 individuals had antibodies to HEV. Seroprevalence increased with age, from 1.1% in children <5 yr old to 14.2% in the group of 26- to 29-yr-olds (Alvarez-Munoz et al. 1999). Similar seroprevalence was observed in Ghana, where 1% of children 6–7 yr of age were HEV seropositive. This number increased to 8.1% among children 16–18 yr of age (Martinson et al. 1999).

F. Enterovirus

In two studies on virus occurrence in solid waste, Peterson (1974) isolated enteroviruses in 10% of the fecally soiled diapers that she examined (Table 9). The excretion rate of enteroviruses has been found to vary with month, with the greatest percentage from May to October in the U.S. The incidence in children over the entire year ranges from 2.4% to 13.3%, with the higher excretion rate in the lower socioeconomic group (Melnick 1997). The most extensive work done on virus excretion was during the "Virus Watch" studies in which the incidence of virus illness and excretion was conducted in families for many years in several locations across the U.S. (Fox and Hall 1980). In Seattle and New York, stool samples were collected from family members regularly, usually at monthly intervals, whether illness was present or not. Over a 3-yr period, the incidence of excretion of any enteric virus in children (<15 yr of age) was found to range from about 10% in the winter to almost 40% during the summer. During summer and autumn months (June through October), more than one-third of healthy children were

Table 10. Incidence of Enteric Virus Infections in the United States.

Pathogen	Incidence (%)	Remarks	Reference
Enterovirus	10	Occurrence in fecally soiled diapers	Peterson 1974
	30–40	During the summer months (June–Oct.): all enteric viruses[a]	Fox and Hall 1980
	2.4–13.3	12 mon average	Melnick 1997
Hepatitis A	0.0097	Reported cases of clinical illness	CDC 1996
	8.2	Occurrence of virus in stools of healthy persons	DeFilippes et al. 1987
Rotavirus	10.4	Annual rate of clinical infection	Ho et al. 1999
	29	All age groups	Champsaur et al. 1984a,b

[a] Any virus isolated from stool samples causing destruction of cell culture.

excreting some virus in the feces, as detected by cell culture. Overall, the frequency of illness associated with echovirus infections was 44%. However, symptomatic infections were greater for children <4 yr of age; 78% for 0- to 4-yr-olds, and 12% for children >5 yr of age and older. The rate of symptomatic infections among adults was 28% for both coxsackievirus and echovirus versus 42% for children <4 y of age (Table 10).

IV. Economic Impact of Enteric Viral Infections in Children
A. Rotavirus

Rotaviruses are the most common cause of severe vomiting and diarrhea in children, with an estimated 3.1 million cases annually in the U.S. (Smith et al. 1995). A study of hospitalizations in children involving diarrhea from 1993–1995 indicated that rotavirus was the most common identified agent (Parashar et al. 1998). In total, viruses accounted for 32.9% of hospitalizations involving diarrhea, followed by bacteria (4.1%), and parasites (0.3%). Overall, rotavirus accounted for 16.5% of hospitalizations for diarrhea among children aged <5 yr. A study by Parashar et al. (1999) on hospital discharge data in Connecticut found that the median cost of diarrhea-associated hospitalization during 1987–1996 and 1993–1996 was $1,941 and $2,428, respectively (Table 11). Because only about half the children admitted to hospitals may be tested, the actual data based on hospital discharges may be underestimated. Hospitalization due to rotavirus gastroenteritis has been estimated at 65,000 to 70,000 annually in the U.S., with 125 deaths (Smith et al. 1995). Ryan et al. (1996) estimated that in England and Wales

Table 11. Indirect and Direct Costs (in Dollars) of Enteric Viral Infections in Children.

Virus	Duration of illness (ds)	Medical costs		Indirect and direct costs		Year[a]	Reference
		Hospitalization	No hospitalization	Hospitalization	No hospitalization		
Rota	3.5	3,615	94	ND	325	1993	Smith et al. 1995
	2	2,428	ND	ND	ND	1993–1996	Parashar et al. 1999
Norwalk	1	887	160–320	1,151	88–480	1993	Crabtree 1996
Hepatitis A	ND	7,138	ND	ND	ND	1999	O'Conner et al. 1999
Enteroviruses							
HFMD	7	ND	69	ND	132	1994	Pichichero et al. 1998
Meningitis	6.5	ND	771	ND	1,193	1994	Pichichero et al. 1998
Echovirus 30 meningitis	ND	1,757	450	ND	ND	1991	Rice et al. 1995
Febrile illness in infants	3	ND	ND	4,500	ND	1997	Byington et al. 199

HFMD, hand-foot-and-mouth disease; ND, no data.
[a]Year for which cost was estimated.

the hospitalization rate for rotavirus-related illness for children <5 yr old was 5/1000.

Smith et al. (1995) estimated, in 1993 dollars, outpatient visits at $94 per visit, hospitalization at $3,615, and loss of productivity at $66/d. For a non-hospitalized child with mild diarrhea (3.5 d of illness), total direct and indirect costs would be $325 per case. Crabtree (1996) estimated total direct and indirect costs of $176/case for those not needing medical attention, $512–$672 for those needing outpatient medical attention, and $4,340 for those needing hospitalization. Yearly costs of childhood rotavirus in the childhood rotavirus in the U.S. were in 1993 determined to be approximately $1.8 billion.

B. Calicivirus

Crabtree (1996) attempted to estimate the cost to individuals due to Norwalk virus illness that might be attributable to drinking water. From a review of outbreaks of Norwalk virus, it was determined that an average of 28% of all cases visited a physician and 2.5% were hospitalized. No deaths were reported in any of the outbreaks; however, to assess costs that may incur from death, a 0.0001% case-fatality rate was applied (Bennett et al. 1987). Direct and indirect costs (1993 dollars) for those who did not see a physician were estimated at $88/case, those seeing a physician at $336–$480/case, and those hospitalized at $1,151/case. It was estimated that more than $1 billion/yr may be associated with Norwalk virus illnesses in the U.S., with nearly $0.5 billion attributable to waterborne transmission (Payment et al. 1991).

C. Hepatitis A Virus

Two studies have examined the cost benefits of hepatitis A immunization in developed countries. O'Connor et al. (1999) evaluated the economic benefits of vaccinated adults and estimated the per case nonhospitalized medical cost at $142, hospitalized at $7,138, and fatal cases at $19,603. Das (1999) estimated average direct medical costs of $1,070–$2,460 per case. Although estimates of indirect costs are available, Lucioni et al. (1998) studied the economic cost of hepatitis A caused specifically by the contamination of food. The direct and indirect cost per patient was $662; costs of hospitalized patients were as great as $86,899.

D. Enterovirus

Studies on the economic cost of non-polio enterovirus infections have only recently been attempted. Pichichero et al. (1998) conducted a study on children >4 yr to assess the economic impact of enterovirus infection. Some 380 children in two clinics, over a period of 4 mon in different regions of the U.S., were involved in the study. The children were followed for 2 wk to document absenteeism and follow-up medical care. The majority of the illnesses were mild, and no hospitalizations were required. Most of the ill-

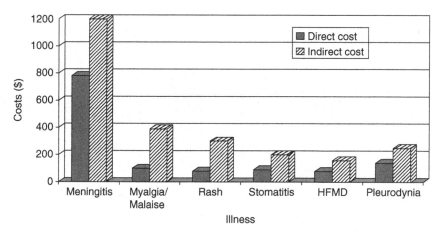

Fig. 4. Comparison of direct and indirect costs of nonpolio enterovirus illness in children. HFMD, hand-foot-and-mouth disease. *Source*: Pichichero et al. (1998).

nesses occurred in children 4–12 yr of age. The duration of illness in most children was prolonged, 9.5 d on average. The total of direct medical care costs and indirect costs per case ranged from $132 for hand-foot-and-mouth disease to $1,193 for meningitis (Fig. 4).

In 1991, a large outbreak of echovirus 30 meningitis occurred in New England, affecting more than 1,500 individuals (Rice et al. 1995). A cost analysis of the hospital billing for the inpatient and outpatient care of 103 patients involved in the outbreak was performed at a hospital serving the region. The average inpatient management cost of a patient with enterovirus meningitis, in this outbreak, was $1,757 ± $198 and the outpatient management cost was $477 ± $63. Indirect costs were not determined. Crabtree (1996) estimated the direct and indirect costs of enterovirus aseptic meningitis to range from $512 to $702 for nonhospitalized cases and $5,403 for hospitalized cases. The indirect costs of cases that did not see a physician were estimated at $176 per case. Bennet et al. (1987) estimated the numbers of aseptic meningitis cases in the U.S. by multiplying the number of estimated enterovirus cases by 6.34%, which is the percent of enterovirus illnesses related to a specific meningitis and reported to the Centers for Disease Control (CDC 1981). From these analyses, the total medical and productivity costs were estimated to range from $1.8 billion to $7 billion annually, with potentially $2.4 billion attributable to water (assuming 35% are waterborne) (Payment et al. 1991).

V. Exposure
A. Drinking Water

Infants and young children have a greater environmental exposure to enteric organisms than adults. They have not yet developed proper sanitary habits (e.g., use of toilet facilities, hand-washing, frequent hand-to-mouth

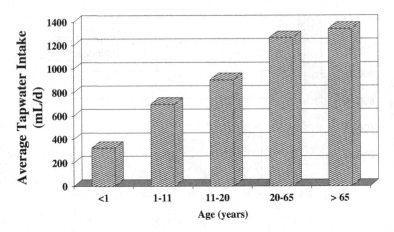

Fig. 5.   Average tapwater intake by age. Data from Roseberry and Burmaster (1992).

or fomite-to-mouth contact) (Springthorpe and Sattar 1990). Object-to-mouth and mouth-to-mouth contact is much greater among children than adults. Viruses are readily transferred from contaminated objects (fomites) directly to the mouth or contamination of the hand. The median number of such contacts by age, per day, are as follows: 1–12 mon, 64; 13–24 mon, 34; 25–30 mon, 27; 31–36 mon, 5; and 37–48 mon, 10 (Hutto et al. 1986). During recreational activities, they may ingest greater quantities of dirt and water (Sedman and Mahmood 1994). However, they do consume less tapwater than adults do (Roseberry and Burmaster 1992), although children <10 yr of age consume more per body weight than any other age group (EPA 2000). Children <11 yr of age consume almost half the amount of tapwater consumed by adults (Fig. 5). However, pregnant and lactating women ingest more tapwater than other women (Burmaster 1998). Because it is believed that many of the serious fetal and neonatal enteric viral infections are contracted from the mother (see enteroviruses in Section C), a greater exposure via tapwater to pregnant and lactating women would imply greater exposure to the fetus and the neonate.

B. Social Economic Factors (Environmental Justice)

Numerous studies have documented the greater incidence of enteric viral infections in lower social economics groups (see II. 3) Behnke et al. 1988. During community-wide outbreaks of hepatitis A in the U.S., neighborhoods of lower socioeconomic status have been identified as a risk factor (Shaw et al. 1986). Households that lack piped water and access to safe water supplies also suffer higher attack rates (Cama et al. 1999; Gurwith et al. 1983; Hyams et al. 1992).

VI. Infectivity (Infectious Dose)

No studies could be found in which an attempt had been made to determine if the infectious dose or infectivity of enteric viruses was different for adults compared to children. Dose–response models have been developed from studies involved in the oral exposure of poliovirus types 1 and 3 (Regli et al. 1991) in which infants and premature babies were used as subjects. The dose response of those viruses is similar to that observed for echovirus 12 and rotavirus in adults; however, infectivity is not directly comparable because this is likely dependent both on the type and strain of virus. Factors that could predispose children to have a greater probability of becoming infected from a given dose than adults are reduced stomach acid and pepsin secretion (Haffejee 1995) (see Section I). Although the severity of illness is usually greater for children than adults, it is currently not known if severity is related to dose for enteric viruses (see II.C. Properties of Enteroviruses).

VII. Risk Assessment

Microbial risk assessment is the application of the principles of risk assessment to estimate the consequences from an exposure to infectious microorganisms. This approach can be used to estimate the magnitude of the risk and the probability of adverse effect (Haas et al. 1999). There is epidemiological evidence that members of all the enteric viruses are transmitted by water (EPA 1999a,b; Haas et al. 1999). Quantitative microbial risk assessments have been conducted for adenovirus (Crabtree et al. 1997), coxsackievirus (Crabtree 1996), and rotavirus (Gerba et al. 1996) in drinking water.

These assessments are patterned after the widely accepted paradigm on chemical risk assessment developed by the National Academy of Sciences in 1983 (NAS 1983). This approach follows the processes of first identifying the pathogen of concern, then developing a dose–response relationship between ingestion of the pathogen and a susceptible host, determining the risk of infection from a given exposure and, finally, characterizing the overall risk.

A. Dose–Response Models

The probability density functions are relatively simple models, with several inherent assumptions (e.g., infection has occurred, the chance of contracting disease is independent of ingested dose). Generally, the first step in the probability analysis is the determination of the probability of infection based on the application of the exponential model or the beta-Poisson model (Haas et al. 1999). The second step is to determine the relationship between infection and developing clinical disease.

In assessing exposure to waterborne adenovirus (Crabtree et al. 1997), an exponential model was used. The probability of becoming infected (P_i) was calculated as:

$$P_i = 1 - \exp(-rN)$$

where

P_i = probability of being infected
N = number of organisms ingested or inhaled
r = 0.4172 for adenovirus (Rose et al. 1996), the probability after ingestion or inhalation that the organism survives to initiate an infectious focus

The "r" value is derived from human exposure studies. For both viruses, the probabilities of developing clinical illness ($P_{illness}$) and of dying as a result of this illness (P_m) were also determined. The probability of becoming ill from exposure was calculated by multiplying the probability of infection (P_i) by the morbidity rate of 0.5 for adenovirus (Haas et al. 1993).

Crabtree et al. (1997) calculated the probability of death by multiplying the probability of infection (P_i) by the mortality rate of 0.0001 for adenovirus.

Gerba et al. (1996) assessed the risks associated with exposure to waterborne rotavirus in tap water using the beta Poisson probability model to calculate probabilities of infection, of illness, and of mortality. The probability of infection was calculated as:

$$P_i = 1 - (1 + N/\beta)^{-\alpha}$$

where

P_i = probability of being infected
β = 0.42 and α = 0.26 (Haas et al. 1993), parameters that describe host-virus interaction after ingestion or inhalation
N = number of organisms ingested

The probability of clinical infection ($P_{illness}$) and the probability of dying as a result of illness (P_m) were also determined. The probability of becoming ill was calculated by multiplying the probability of infection (P_i) by the morbidity rate of 0.5 (Haas et al. 1993). The probability of death (mortality) from infection was calculated by multiplying the probability of infection (P_i) by the morbidity rate and a case/fatality of 0.0001.

The beta Poisson model has been identified as the model that best fits most of the dose–response data for viruses and which provides a conservative method for low-dose extrapolation (Haas et al. 1993; Regli et al. 1991). However, several assumptions are made that limit the use of probability models from estimating risks from exposure to pathogens in drinking water. The Poisson model assumes random distribution of microorganisms in drinking water. The risk estimates for low-level exposures are based on extrapolation to low doses and, at very low pathogen concentration, the relationship between risk of infection and dose is approximately linear (Haas et al. 1993). The models assume that the exposed population is equally susceptible to a single exposure and ingests 2 L of water per day. The relationship between infection and the development of clinical illness is regarded as a conditional probability that, once having been infected, a certain number of individuals will develop disease (Haas et al. 1993). The chance of developing a symptomatic illness once infected by a virus is assumed to be independent of dose. Each exposure is regarded as statistically independent [i.e., the chance of developing an infection (illness or death) from one exposure is not related to prior exposures and effects]. The calculated risk is for the nonimmune person; therefore, immunity plays no role in the risk assessment for a nonimmune person.

B. Epidemiological Evidence for Transmission of Viral Diseases to Children by Water

Because of the need to consume fluids, drinking water-associated outbreaks inadvertently affect all age groups. Two epidemiological studies, designed to study the impact of conventionally treated drinking water meeting all standards, found that children were the most affected group (Payment et al. 1991). Figure 6 shows the relative incidence of gastroenteritis by age in those individuals drinking tapwater and those drinking tapwater after filtration by a reverse-osmosis filter, designed to remove pathogens (Payment et al. 1991). A greater impact of highly credible gastrointestinal illness was

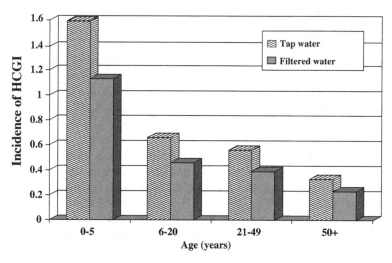

Fig. 6. Incidence of highly credible gastrointestinal symptoms (HCGI) episodes for period 2 by selected population subgroups. Data from Payment et al. (1991).

seen among children who drank the nonfiltered tapwater. In the other study, families were provided with purified bottled water, or with tapwater delivered from the water treatment plant after treatment (bottled tapwater), or drank only tapwater after it had been delivered through the distribution system (Payment et al. 1997). Using the purified water as the baseline, the excess of gastrointestinal illness associated with tapwater was 14% in the bottled tapwater group and 19% in those who consumed water from the tap in the home. Children 2–5 yr old were the most affected, with an excess of 17% in the bottle tapwater group and 40% in those drinking water from the tap. In either study, the agent causing the illness was not identified. The system was served by a poor-quality surface water.

Rotavirus. Rotavirus has been responsible for several drinking waterborne outbreaks worldwide (Gerba et al. 1996). In a 1981 outbreak of rotavirus in Colorado, an estimated 1,500 individuals were infected, including both adults and children (Hopkins et al. 1984). In a waterborne outbreak in a school in Brazil, higher attack rates of gastroenteritis were seen in nursery-age and kindergarten-age children (Sutmoller et al. 1982). A study in Lima, Peru, found that attack rates were higher in children who were not exclusively breast-fed in early infancy and who also lacked piped water in their homes (Cama et al. 1999).

Caliciviruses. Numerous drinking waterborne outbreaks of caliciviruses have been documented. Several outbreaks have occurred at elementary schools and summer camps (Kaplan et al. 1982; Taylor et al. 1981), while others were community outbreaks in which individuals of all ages were affected (Goodman et al. 1982). Although attack rates were similar for both adults and children, secondary transmission rates were greater among children. In one swimming-associated outbreak, the secondary attack rate was highest among children <10 yr of age (Baron et al. 1982). Also, attack rates were significantly higher during common source outbreaks, such as drinking water (median, 60% vs. 39% for person-to-person) (Kaplan et al. 1982). A study of viral diarrhea in three native Indian villages in Canada noted that infections of Norwalk virus were greatest in infants in the one community with an untreated water supply (Gurwith et al. 1983).

Adenovirus. Although there has only been two suspected drinking water outbreak involving an adenovirus (Papapetropoulou and Vantarakis 1998), there have been numerous outbreaks of swimming-associated adenovirus infections, many involving children (Foy et al. 1968; McMillian et al. 1992; Martone et al. 1980). Outbreaks have been associated with adenovirus 3, 4, and 7 causing conjunctivitis or pharyngoconjunctival fever affecting children 1–18 yr of age. Attack rates have been as high as 67% in children, with secondary attack rates of 19% for adults and 63% for children (Foy et al. 1968) (Fig. 7).

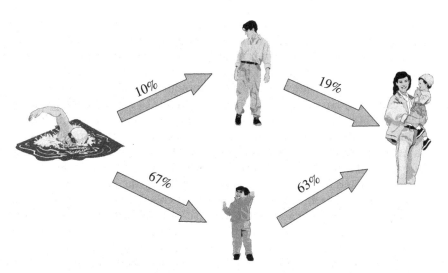

Fig. 7. Impact of a waterborne disease outbreak of adenovirus on attack rates and person-to-person transmission.

Hepatitis A Virus (HAV).   Waterborne outbreaks of HAV are well documented in the U.S. (Gerba et al. 1985). Although asymptomatic infections are more common among children than adults (see earlier section), higher attack rates have been observed during waterborne outbreaks in older children. In an outbreak in Maryland traced to a heavily contaminated spring used as drinking water, the highest attack rates were in the 10–14-yr age group (Whatley et al. 1968). In an outbreak associated with a water fountain, the 16–20-yr age group (Bowen and McCarthy 1983) was affected the most.

Hepatitis E Virus (HEV).   During the past decade, at least 30 outbreaks of HEV have been associated with waterborne outbreaks involving drinking water in 17 countries (Craske 1992). Although outbreaks of hepatitis E have not been reported in the U.S., it has recently been found to occur in pigs, which may serve as a reservoir of human infection (Meng et al. 1998, 1999). Although children become infected with HEV during waterborne outbreaks, the most serious resulting illness occurs in adults. In two drinking water-associated outbreaks in Mexico, the attack rate for persons <15 yr of age was 1%–2% vs. 10% for those >15 yr of age. However, mortality among pregnant women generally ranges from 20% to 30% and can be as high as 40% (Craske 1992). Thus, the fetus is at serious risk of mortality during waterborne outbreaks. Hepatitis E infection in children has also been associated with the lack of indoor plumbing in the developing world (Hyams et al. 1992).

Enteroviruses. Although there have been no clearly documented drinking water outbreaks associated with enteroviruses, several recreational outbreaks have been documented among children. An outbreak of coxsackievirus B4 or B5 meningitis linked to swimming in a lake occurred at a boys' summer camp (Hawley et al. 1973). Another outbreak of meningitis, this time caused by coxsackievirus A16, occurred due to exposure to lake water (Denis 1974). An epidemiological study among bathers swimming in nondisinfected lake waters demonstrated an association with increased risk of enterovirus infection. D'Alessio et al. (1981) surveyed children 1–15 yr of age at a pediatric clinic to determine where they had been swimming and the location and frequency of the swimming during the prior 2 wk. Children swimming at a beach versus in a pool with a chlorine residual had a statistically significant increase in the relative risk of having an enterovirus illness. Of the 134 viruses isolated from the patients, 119 (90%) were non-polio enteroviruses and 33.6% were coxsackievirus type A.

Echovirus 30 transmission to children has also been associated with a community swimming pool (Kee et al. 1994). The risk of echovirus 30 was greatest among those who swallowed pool water.

C. Endpoints

An assessment endpoint for microbial risk assessment has usually been risk of infection (Regli et al. 1991). Greater uncertainty exists in assessing the probability of illness and mortality, because this is dependent not only on the type of virus but on also a particular strain. Other factors include the immune state of the host, age, and other preexisting conditions. In the case of HAV, very young children are more likely to develop clinical symptoms than older children and adults, and the disease is generally more severe in adults. However, in the case of rotavirus, the resulting diarrhea is more severe in children than adults, which is reflected by the large number of hospitalizations for rotavirus infections in the U.S. Adenoviruses and astroviruses appear to be largely involved in infections of children. In contrast, Norwalk viruses appear to affect all age groups almost equally. Enteroviruses cause a wide range of illnesses, being most severe in neonates and children. Although most children recover, infections in neonates, especially of coxsackieviruses, are frequently fatal. With the possible exceptions of hepatitis A and E, a greater severity of illness and risk of mortality exists for children than adults. Thus, in assessing the risks of enteric viral infections in children, it is important to assess the endpoints of severity of illness and mortality because they can be significantly greater than in adults.

D. Risk Characterization of Enteric Viruses in Water and Children

The only dose–response data available for children are those obtained from studies conducted with vaccine strains of poliovirus. Lepow et al. (1962) conducted studies on newborn infants less than 5 d old with poliovirus Sabin

Table 12. Probability of Infection from Different Types of Poliovirus by the Oral Route in Children.

Virus type and Strain	Probability of infection	ID_{50}[a]	Reference
1 LSc-2	7.14×10^{-4}	6.93×10^4	Lepow et al. 1962
1	9.10×10^{-3}	76.2	Minor et al. 1981
3 Fox	1.90×10^{-1}	5.5	Plotkin et al. 1959
3 Fox	2.66×10^{-1}	5.0	Katz and Plotkin 1967

[a]The number of viruses required to cause infection in 50% of the individuals exposed.
Source: Teunis et al. (1996).

Table 13. Probability of Infection by Poliovirus Type 1 by Age-Adjusted Exposures for Tapwater Ingestion.[a]

Age (yr)	Exposure (L)	Daily risks of infection	Yearly risks of infection
<1	0.323	2.94×10^{-4}	1.04×10^{-1}
1–11	0.701	6.38×10^{-4}	2.08×10^{-1}
11–20	0.907	8.25×10^{-4}	2.60×10^{-1}
21–65	1.265	1.15×10^{-3}	3.43×10^{-1}

[a]Assuming ingestion of one infectious virus in 10 L of drinking water.
Exponential model of Teunis et al. (1996); $r = 0.009102$.

type 1 (LSc-2). Minor et al. (1980) exposed 2-mon-old infants via the oral route with a syringe with a vaccine strain of polio type 1. Plotkin et al. (1959) and Katz and Plotkin (1967) orally exposed premature infants to an attenuated strain of poliovirus type 3 (Fox strain). The probability of being infected by ingesting one virus in these studies was assessed by Teunis et al. (1996). They found that the probability ranged from 7.14×10^{-4} to 1.90×10^{-1} (Table 12). The range probably reflects the type of virus and method of administration. Using the data of Roseberry and Burnmaster (1992) on average ingestion of tapwater by age, an assessment was made of the probability of infection in different age groups for poliovirus type 1 (Table 13).

The risk of infection, illness, and death associated with coxsackievirus levels in conventionally treated water are shown in Table 14 for children. The data suggest that significant risks of illness for children could exist from these exposure levels; however, outcome would always be dependent on the virulence of the individual virus.

VIII. Conclusions

This review suggests that children and immunocompromised individuals bear the greatest burden of illness associated with drinking water-transmitted enteric viral diseases. They suffer the highest attack rates and

Table 14. Risk of Infection, Illness, and Death Associated with Coxsackievirus Level in Tapwater.

Age	Exposure (L)	Daily risk			Yearly risk		
		Infection	Illness	Mortality	Infection	Illness	Mortality
<1	0.323	1.2×10^{-5}	9.34×10^{-6}	5.6×10^{-8}	4.5×10^{-3}	3.4×10^{-3}	$1.99 \times 10^-$
1–11	0.701	2.7×10^{-5}	2.0×10^{-5}	1.2×10^{-7}	9.8×10^{-3}	7.3×10^{-3}	$4.3 \times 10^-$

[a]Concentration of 0.005 MPNCU (most probable number of cytopathic units)/L of virus in the tapwater (Payment et al. 1985): surface water supply.
Risk of infection and mortality from Crabtree (1996).
A cytopathic unit is defined as destruction of cells in culture by a virus.

the most severe illness. To better quantify the impact on children, the literature should be further reviewed for case studies of waterborne outbreaks where data are available on the resulting illness by age group. The EPA and/or Centers for Disease Control should attempt to collect these data as future outbreaks are documented.

Given the differences in the physiology between children and adults, it may be that children have a greater probability of infection with a given dose than adults. Studies in animals or vaccine strains of viruses with children should be conducted to determine if a significant difference exists. Because the major route of infection of neonates by enteroviruses appears to be transmission from the mother to the child, at or shortly after birth, greater development of dose–response data in animals may be useful to assess if a greater susceptibility of the mother to enteroviruses occurs during pregnancy. Better documentation of long-term sequelae, particularly for enterovirus infections, is needed.

Summary

Children are at a greater risk of infections from serious enteric viral illness than adults for a number of reasons. Most important is the immune system, which is needed to control the infection processes. This difference can lead to more serious infections than in adults, who have fully developed immune systems. There are a number of significant physiological and behavioral differences between adults and children that place children at a greater risk of exposure and a greater risk of serious infection from enteric viruses.

Although most enteric viruses cause mild or asymptomatic infections, they can cause a wide range of serious and life-threatening illnesses in children. The peak incidence of most enteric viral illnesses is in children <2 yr of age, although all age groups of children are affected. Most of these infections are more serious and result in higher mortality in children than adults. The fetus is also affected by enterovirus and infectious hepatitis resulting in significant risk of fetal death or serious illness. In addition to the poliovirus vaccine, the only vaccine available is for hepatitis A virus (HAV). A vaccine for rotavirus has currently been withdrawn, pending review because of potential adverse effects in infants. No specific treatment is available for the other enteric viruses.

Enteric viral infections are very common in childhood. Most children are infected with rotavirus during the first 2 yr of life. The incidence of enteroviruses and the viral enteric viruses ranges from 10% to 40% in children and is largely dependent on age. On average, half or more of the infections are asymptomatic. The incidence of hepatitis A virus is much lower than the enteric diarrheal viruses. There is no current evidence for hepatitis E virus (HEV) acquisition in children in the U.S.

Enteric viral diseases have a major impact on direct and indirect health care costs (i.e., lost wages) and amount to several billion dollars a year in the U.S. Total direct and indirect costs for nonhospitalized cases may run from $88/case for Norwalk virus to $1,193/case for enterovirus aseptic meningitis. Direct costs of hospitalization ran from $887/case for Norwalk virus to $86,899/case for hepatitis A. These costs are based on 1997–1999 data.

Generally, attack rates during drinking water outbreaks are greater for children than adults. The exception appears to be hepatitis E virus where young adults are more affected. However, pregnant women suffer a high mortality, resulting in concurrent fetal death. Also, secondary attack rates are much higher among children, probably because of fewer sanitary habits among this age group. Overall, waterborne outbreaks of viral disease have a greater impact among children than adults.

To better quantify the impact on children, the literature hould be further reviewed for case studies of waterborne outbreaks where data are available on the resulting illness by age group. The EPA and/or Centers for Disease Control should attempt to collect these data as future outbreaks are documented.

Acknowledgments

We acknowledge Arthur Reingold, Walter Jakubowski, and Ricardo Deleon for reviewing this manuscript.

References

Abramson, JS, Baker, CJ, Fisher, MC, Gerber, MA, Meissner, HC, Murray, DL, Overturf, GD, Prober, CG, Rennels, MB, Saari, TN, Weiner, LB, and Whitley, RJ (1999) Possible association of intussusception with rotavirus vaccination. American Academy of Pediatrics. Committee on Infectious Diseases. Pediatrics 104:575.

Abzug, MJ, Keyserling, HL, Lee, ML, Levin, MJ, and Rotbart, HA (1995) Neonatal enterovirus infection: virology, serology, and effects of intravenous immune globulin. Clin Infect Dis 20:1201–1206.

Albert, MJ (1986) Enteric adenoviruses. Brief review. Arch Virol 88:1–17.

Alvarez-Munoz, MJ, Torres, J, Damasio, L, Gomez, A, Tapia-Conyer, R, and Munoz, O (1999) Seroepidemiology of hepatitis E virus infection in Mexican subjects 1 to 29 years of age. Arch Med Res 30:251–254.

Appleton, H (1987) Small round viruses: classification and role in food-borne infections. Ciba Found Symp 128:108–125.

Appleton, H, and Pereira, MS (1977) A possible virus aetiology in outbreaks of food-poisoning from cockles. Lancet 1(8015):780–781.

Aquino, VM, Farah, RA, Lee, MC, and Sandler, ES (1996) Disseminated coxsackie A9 infection complicating bone marrow transplantation. Pediatr Infect Dis J 15:1053–1054.

Armstrong, GL, and Bell, BP (2002) Hepatitis A virus infections in the United States: model-based estimates and implications for childhood immunizations. Pediatrics 109:839–845.

Arora, NK, Nanda, SK, Gulati, S, Ansari, IH, Chawla, MK, Gupta, SD, and Panda, SK (1996) Acute viral hepatitis types E, A, and B singly and in combination in acute liver failure in children in north India. J Med Virol 48:215–221.

Assaad, F, and Borecka, I (1977) Nine-year study of WHO virus reports on fatal viral infections. Bull WHO 55:445–453.

Balayan, MS (1993) Hepatitis E virus infection in Europe: regional situation regarding laboratory diagnosis and epidemiology. Clin Diagn Virol 1:1–9.

Ball, AP (1975) Disease due to echovirus type 19 in Birmingham, England, 1975: Relationship to "epidemic neuromyasthenia." Postgrad Med J 54:737–740.

Barnes, GL, Uren, E, Stevens, KB, and Bishop, RF (1998) Etiology of acute gastroenteritis in hospitalized children in Melbourne, Australia, from April 1980 to March 1993. J Clin Microbiol 36:133–138.

Baron, RC, Murphy, FD, Greenberg, HB, Davis, CE, Bregman, DJ, Gary, GW, Hughes, JM, and Schonberger, LB (1982) Norwalk gastrointestinal illness: an outbreak associated with swimming in a recreational lake and secondary person-to-person transmission. Am J Epidemiol 115:163–172.

Bartlett, AV III, Reves, RR, and Pickering, LK (1988) Rotavirus in infant-toddler day care centers: Epidemiology relevant to disease control strategies. J Pediatr 113:435–441.

Behnke, M, Davis-Eyler, F, Conlon, M, Quiros-Casanova, O, and Stewart-Woods, N (1997) How fetal cocaine exposure increases neonatal hospital costs. Pediatrics 99:204–208.

Bellanti, JA, Nerurkar, LS, and Zeligs, BJ (1979) Host defenses in the fetus and neonate: studies of the alveolar macrophage during maturation. Pediatrics 64:726–739.

Beneson, AS (1990) Control of communicable diseases in man, 15th Ed. American Public Health Association, Washington, DC.

Bennett, JV, Holmberg, SD, Rogers, MF, and Solomon, SL (1987) Infectious and parasitic diseases. Am J Prev Med 102:S3–S5.

Berge, JJ, Drennan, DP, Jacobs, RJ, Jakins, A, Meyerhoff, AS, Stubblefield, W, and Weinberg, M (2000) The cost of hepatitis A infections in American adolescents and adults in 1997. Hepatology 31:469–473.

Berger, MM, Kopp, N, Vital, C, Redl, B, Aymard, M, and Lina, B (2000) Detection and cellular localization of enterovirus RNA sequences in spinal cord of patients with ALS. Neurology 54:20–25.

Berke, T, and Matson, DO (2000) Reclassification of the Caliciviridae into distinct genera and exclusion of hepatitis E virus from the family on the basis of comparative phylogenetic analysis. Arch Virol 145:1421–1436.

Birenbaum, E, Handsher, R, Kuint, J, Dagan, R, Raichman, B, Mendelson, E, and Linder, N (1997) Echovirus type 22 outbreak associated with gastro-intestinal disease in a neonatal intensive care unit. Am J Perinatol 14:469–473.

Blacklow, NR, and Greenberg, HB (1991) Viral gastroenteritis. N Engl J Med 325:252–264.

Blacklow, NR, Herrmann, JE, and Cubitt, WD (1987) Immunobiology of Norwalk virus. Ciba Found Symp 128:144–161.

Blotzer, JW, and Myers, AR (1978) Echovirus-associated polyarthritis. Report of a case with synovial fluid and synovial histologic characterization. Arth Rheum 21:978–981.

Bowen, GS, and McCarthy, MA (1983) Hepatitis A associated with a hardware store water fountain and a contaminated well in Lancaster County, Pennsylvania, 1980. Am J Epidemiol 117:695–705.

Boyd, MT, Jordan, SW, and Davis, LE (1987) Fatal pneumonitis from congenital echovirus type 6 infection. Pediatr Infect Dis J 6:1138–1139.

Bradley, DW, Beach, MJ, and Purdy, MA (1992) Recent developments in the molecular cloning and characterization of hepatitis C and E viruses. Microb Pathog 12:391–398.

Brandt, CD, Kim, HW, Yolken, RH, Kapikian, AZ, Arrobio, JO, Rodriguez, WJ, Wyatt, RG, Chanock, RM, and Parrott, RH (1979) Comparative epidemiology of two rotavirus serotypes and other viral agents associated with pediatric gastroenteritis. Am J Epidemiol 110:243–254.

Brandt, CD, Kim, HW, Rodriguez, WJ, Arrobio, JO, Jeffries, BC, Stallings, EP, Lewis, C, Miles, AJ, Gardner, MK, and Parrott, RH (1985) Adenoviruses and pediatric gastroenteritis. J Infect Dis 151:437–443.

Brown, GC, and Karunas, RS (1971) Relationship of congenital anomalies and maternal infection with selected enteroviruses. Am J Epidemiol 95:207–217.

Burmaster, DE (1998) Lognormal distributions for total water intake and tap water intake by pregnant and lactating women in the United States. Risk Anal 18:215–219.

Butz, AM, Fosarelli, P, Dick, J, Cusack, T, and Yolken, R (1993) Prevalence of rotavirus on high-risk fomites in day-care facilities. Pediatrics 92:202–205.

Byington, CL, Taggart, EW, Carroll, KC, and Hillyard, DR (1999) A polymerase chain reaction-based epidemiologic investigation of the incidence of nonpolio enteroviral infections in febrile and afebrile infants 90 days and younger. Pediatrics 103:E27.

Calisher, CH, and Fauquet, CM (1992) Stedman's ICTV virus words. In: International Committee on Taxonomy of Viruses. Williams & Williams, Baltimore.

Cama, RI, Parashar, UD, Taylor, DN, Hickey, T, Figueroa, D, Ortega, YR, Romero, S, Perez, J, Sterling, CR, Gentsch, JR, Gilman, RH, and Glass, RI (1999) Enteropathogens and other factors associated with severe disease in children with acute watery diarrhea in Lima, Peru. J Infect Dis 179:1139–1144.

Caul, EO (1996a) Viral gastroenteritis: small round structured viruses, caliciviruses and astroviruses. Part I. The clinical and diagnostic perspective. J Clin Pathol 49:874–880.

Caul, EO (1996b) Viral gastroenteritis: small round structured viruses, caliciviruses and astroviruses. Part II. The epidemiological perspective. J Clin Pathol 49:959–964.

CDC (Centers for Disease Control) (1981) Aseptic meningitis in a high school football team – Ohio. Morb Mortal Wkly Rep 30:631.

CDC (1996) Prevention of hepatitis A through active or passive immunization: Recommendations of the advisory committee on immunization practices. Morb Mortal Wkly Rep 45:1–30.

CDC (1999a) Intussusception among recipients of rotavirus vaccine – United States, 1998–1999. Morb Mortal Wkly Rep 48:577–581.

CDC (1999b) Prevention of hepatitis A through active or passive immunization: recommendations of the Advisory Committee on Immunization Practices (ACIP). Morb Mortal Wkly Rep 48:1–37.

CDC (1999c) Rotavirus vaccine for the prevention of rotavirus gastroenteritis among children: Recommendations of the Advisory Committee on Immunization Practices (ACIP). Morb Mortal Wkly Rep 48:1–23.

Champsaur, H, Henry-Amar, M, Goldszmidt, D, Prevot, J, Bourjouane, M, Questiaux, E, and Bach, C (1984a) Rotavirus carriage, asymptomatic infection, and disease in the first two years of life. II. Serological response. J Infect Dis 149:675–682.

Champsaur, H, Questiaux, E, Prevot, J, Henry-Amar, M, Goldszmidt, D, Bourjouane, M, and Bach, C (1984b) Rotavirus carriage, asymptomatic infection, and disease in the first two years of life. I. Virus shedding. J Infect Dis 149:667–674.

Cheeseman, SH, Hirsch, MS, Keller, EW, and Keim, DE (1977) Fatal neonatal pneumonia caused by echovirus type 9. Am J Dis Child 131:1169.

Chonmaitree, T, and Mann, L (1995) Respiratory infections. In: Rotbart, HA (ed) Human Enterovirus Infections. ASM Press, Washington, DC, pp 255–270.

Chuang, E, Maller, ES, Hoffman, MA, Hodinka, RL, and Altschuler, SM (1993) Successful treatment of fulminant echovirus 11 infection in a neonate by orthotopic liver transplantation. J Pediatr Gastroenterol Nutr 17:211–214.

Crabtree, KD (1996) Risk assessment of viruses in water. Ph.D. dissertation, University of Arizona, Tucson.

Crabtree, KD, Gerba, CP, Rose, JN, and Haas, CN (1997) Waterborne adenovirus: A risk assessment. Water Sci Technol 35:1–6.

Craske, J (1992) Hepatitis C and non-A non-B hepatitis revisited: hepatitis E, F and G. J Infect 25:243–250.

Craver, RD, and Gohd, R (1990) Fatal pneumonitis caused by echovirus 17. Pediatr Infect Dis J 9:453–454.

Crowley, DS, Ryan, MJ, and Wall, PG (1997) Gastroenteritis in children under 5 years of age in England and Wales. Communi Dis Rep CDR Rev 7:R82–R86.

Cruz, JR, Caceres, P, Cano, F, Flores, J, Bartlett, A, and Torun, B (1990) Adenovirus type 40 and 41 and rotaviruses associated with diarrhea in children from Guatamala. J Clin Microbiol 28:1780–1784.

Cubitt, WD (1987) The candidate caliciviruses. Ciba Found Symp 128:126–143.

Cummins, AG, Eglinton, BA, Gonzalez, A, and Roberton, DM (1994) Immune activation during infancy in healthy humans. J Clin Immunol 14:107–115.

Dagan, R. (1996) Nonpolio enteroviruses and the febrile young infant: epidemiologic, clinical and diagnostic aspects. Pediatr Infect Dis J 15:67–71.

Dagan, R, Hall, CB, Powell, KR, and Menegus, MA (1989) Epidemiology and laboratory diagnosis of infection with viral and bacterial pathogens in infants hospitalized for suspected sepsis. J Pediatr 115:351–356.

D'Alessio, D, Minor, TE, Allen, CI, Tsiatis, AA, and Nelson, DB (1981) A study of the proportions of swimmers among well controls and children with enterovirus-like illness shedding or not shedding an enterovirus. Am J Epidemiol 113:533–541.

D'Angelo, LJ, Hierholzer, JC, Keenlyside, RA, Anderson, LJ, and Martone, WJ (1979) Pharyngoconjunctival fever caused by adenovirus type 4: report of a swimming pool-related outbreak with recovery of virus from pool water. J Infect Dis 140:42–47.

Das, A (1999) An economic analysis of different strategies of immunization against hepatitis A virus in developed countries [see comments]. Hepatology 29:548–552.

De Filippis, P, Divizia, M, Mele, A, Adamo, B, and Pana, A (1987) Detection of hepatitis A virus in the stools of healthy people from endemic areas. Eur J Epidemiol 3:172–175.

De Jong, JC, Wermenbol, AG, Verweij-Uijterwaal, MW, Slaterus, KW, Wertheim-Van Dillen, P, Van Doornum, GJ, Khoo, SH, and Hierholzer, JC (1999) Adenoviruses from human immunodeficiency virus-infected individuals, including two strains that represent new candidate serotypes Ad50 and Ad51 of species B1 and D, respectively. J Clin Microbiol 37:3940–3945.

Deneen, VC, Hunt, JM, Paule, CR, James, RI, Johnson, RG, Raymond, MJ, and Hedberg, CW (2000) The impact of fooodborne calicivirus disease: the Minnesota experience. J Infect Dis 181(suppl 2):S281–S283.

Denis, FA, Blanchouin, E, Lignieres, AD, and Flamen, P (1974) Letter: Coxsackie A16 infection from lake water. JAMA 228:1370–1371.

Dery, P, Marks, MI, and Shapera, R. (1974) Clinical manifestations of coxsackievirus infections in children. Am J Dis Child 128:464–468.

Dewey, KG, Heinig, MJ, and Nommsen-Rivers, LA (1995) Differences in morbidity between breast-fed and formula-fed infants. Pediatrics 126:696–702.

Divizia, M, Gabrieli, R, Donia, D, Macaluso, A, Bosch, A, Guix, S, Sanchez, Villena, C, Pinto, RM, Palombi, L, Buonuomo, E, Cenko, E, Leno, L, Bebeci, and Bino, S (2004) Waterborne gastroenteritis outbreak in Albania. Water Sci Technol 50:57–61.

Eichenwald, HF, McCracken, JGH, and Kindberg, SJ (1967) Virus infections of the newborn. Prog Med Virol 9:35–104.

el-Sageyer, MM, Szendroi, A, Hutter, E, Uj, M, Szucs, G, Mezey, I, Toth, I, Katai, A, Kapiller, Z, Pall, G, Petras, G, Szalay, E, Mihaly, I, Gourova, S, and Berencsi, G (1998) Characterisation of an echovirus type 11' (prime) epidemic strain causing haemorrhagic syndrome in newborn babies in Hungary. Acta Virol 42:157–166.

Enriquez, CE, Hurst, CJ, and Gerba, CP (1995) Survival of the enteric adenovirus-40 and adenovirus-41 in tap, sea, and waste-water. Water Res 29:2548–2553.

EPA (1999a) Drinking Water Criteria Document for Viruses: An Addendum. Office of Water, Washington, DC.

EPA (1999b) Drinking water criteria document for enteroviruses and hepatitis A: An addendum. Office of Water, Washington, DC.

EPA (2000) Estimated per capita water injestion in the United States. Office of Water, Washington, DC.

Estes, MK, and Hardy, ME (1995) Norwalk virus and other enteric calicivirus. Infections of the Gastrointestinal Tract. Raven Press, New York.

Etter, CG, Wedgwood, J, and Schaad, UB (1991) Aseptische Meningitiden in der Padiatrie. Schweiz Med Wochenschr J Suisse Med 121:1120–1126.

Evans, HS, Madden, P, Douglas, C, Adak, GK, O'Brien, SJ, Djuretic, T, Wall, PG, and Stanwell-Smith, R (1998) General Outbreaks of infectious intestinal disease in England and Wales; 1995 and 1996. Commun Dis Public Health 1:165–171.

Ferson, MJ (1996) Hospitalisations for rotavirus gastroenteritis among children under five years of age in New South Wales. Med J Aust 164:273–277.

Fischer, TK, Bresee, JS, and Glass, RI (2004) Rotavirus vaccines and the prevention of hospital-acquired diarrhea in children. Vaccine 22(suppl 1):S49–S54.

Fishman, LN, Jonas, MM, and Lavine, JE (1996) Update on viral hepatitis in children. Pediatr Clin N Am 43:57–74.

Forbes, JA (1963) Some clinical aspects of meningoencephalitis. Med J Aust 1:568–572.

Fox, JP, and Hall, CE (1980) Viruses in families. PSG Publishing, Littleton, MA.

Foy, HM, Cooney, MK, and Hatlen, JB (1968) Adenovirus type 3 epidemic associated with intermittent chlorination of a swimming pool. Arch Environ Health 17:795–802.

Galbraith, DN, Nairn, C, and Clements, GB (1997) Evidence for enteroviral persistence in humans. J Gen Virol 78:307–312.

Gauntt, CJ, Gudvangen, RJ, Brans, YW, and Marlin, AE (1985) Coxsackievirus group B antibodies in the ventricular fluid of infants with severe anatomic defects in the central nervous system. Pediatrics 76:64–68.

Gear, JHS, and Measroch, V (1973) Coxsackievirus infections of the newborn. Prog Med Virol 15:42–62.

Geller, TJ, and Condie, D (1995) A case of protracted coxsackie virus meningoencephalitis in a marginally immunodeficient child treated successfully with intravenous immunoglobulin. J Neurol Sci 129:131–133.

Gerba, CP, Rose, JB, and Singh, SN (1985) Waterborne gastroenteritis and viral hepatitis. CRC Crit Rev Environ Control 15:213–236.

Gerba, CP, Rose, JB, and Haas, CN (1996) Waterborne rotavirus: risk assessment. Water Res 30:2929–2940.

Gerba, CP, Rose, JB, and Haas, CN (1996) Sensitive populations: Who is at the greatest risk? Int J Food Microbiol 30:113–123.

Gerba, CP, Enriquez, CE, and Nwachuku, N (2000) Health risks of waterborne enteric viral infections in children. In: Abstracts of the American Society for Microbiology 100th General Meeting, Los Angeles, CA, May 21–25, 2000, p 604.

Gewurz, A, Potempa, R, Goetz, C (1985) Coxsackie A-11 encephalitis (CAE) in a patient with common variable immunodeficiency (CVID): Response to intravenous and intraventricular treatment with intravenous immune globulin (IVIG). Ann Allergy 55:272.

Girones, R, Allard, A, Wadell, G, and Jofre, V (1993) Application of PCR to the detection of adenoviruses in polluted waters. Water Sci Technol 27:235–241.

Glass, RI, Noel, J, Ando, T, Fankhauser, R, Belliot, G, Mounts, A, Parashar, UD, Bresee, JS, and Monroe, S (2000) The epidemiology of enteric caliciviruses from humans: A reassessment using new diagnostics. J Infect Dis 181(suppl 2):S254–S261.

Goodman, RA, Buehler, JW, Greenberg, HB, McKinley, TW, and Smith, JD (1982) Norwalk gastroenteritis associated with a water system in a rural Georgia community. Arch Environ Health 37:358–360.

Gouvea, V, Santos, N, Timenetsky, MDC, and Estes, MK (1994) Identification of Norwalk virus in artificially seeded shellfish and selected foods. J Virol Methods 48:177–187.

Green, KY, Ando, T, Balayan, MS, Berke, T, Clarke, IN, Estes, MK, Matson, DO, Nakata, S, Neill, JD, Studdert, MJ, and Thiel, H-J. (2000) Taxonomy of the caliciviruses. J Infect Dis 181(suppl 2):S322–S330.

Green, J, Casabonne, D, and Newton, R (2004) Coxsackie B virus serology and type 1 diabetes mellitus: A systematic review of published case-control studies. Diabetes Med 21:503–506.

Grist, NR, and Bell, EJ (1970) Enteroviral etiology of the paralytic poliomyelitis syndrome. Arch Environ Health 21:382–387.

Grist, NR, and Bell, EJ (1984) Paralytic poliomyelitis and nonpolio enteroviruses: studies in Scotland. Rev Infect Dis 6(suppl 2):S385–S386.

Gurwith, M, Wenman, W, Gurwith, D, Brunton, J, Feltham, S, and Greenberg, H (1983) Diarrhea among infants and young children in Canada: A longitudinal study in three northern communities. J Infect Dis 147:685–692.

Haas, CN, Rose, JB, Gerba, C, and Regli, S (1993) Risk assessment of virus in drinking water. Risk Anal 13:545–552.

Haas, CN, Rose, JB, and Gerba, CP (1999) Quantitative Microbial Risk Assessment. Wiley, New York.

Haffejee, IE (1995) The epidemiology of rotavirus infections: a global perspective. J Pediatr Gastroenterol Nutr 20:275–286.

Haffejee, IE, Moosa, A, and Windsor, I (1990) Circulating and breast-milk anti-rotaviral antibodies and neonatal rotavirus infections: a maternal-neonatal study. Ann Trop Paediatr 10:3–14.

Hagiwara, A, Tagaya, I, and Yoneyama, T (1978) Epidemic of hand, foot and mouth disease associated with enterovirus 71 infection. Intervirology 9:60–63.

Harrison, TJ (1999) Hepatitis E virus – an update. Liver 19:171–176.

Hawley, HB, Morin, DP, Geraghty, ME, Tomkow, J, and Phillips, CA (1973) Coxsackievirus B epidemic at a Boy's Summer Camp. Isolation of virus from swimming water. JAMA 226:33–36.

Heaton, DC, and Moller, PW (1985) Still's disease associated with Coxsackie infection and haemophagocytic syndrome. Ann Rheum Dis 44:341–344.

Heinberg, H, Gold, E, and Robbins, FC (1964) Difference in interferon content in tissue of mice of various ages infected with coxsackie B-1 virus. Proc Soc Exp Biol Med 115:947–953.

Helfand, RF, Khan, AS, Pallansch, MA, Alexander, JP, Meyers, HB, DeSantis, RA, Schonberger, LB, and Anderson, LJ (1994) Echovirus 30 infection and aseptic meningitis in parents of children attending a child care center. J Infect Dis 169:1133–1137.

Herrmann, JE, and Blacklow, NR (1995) Enteric adenoviruses. Infections of the Gastrointestinal Tract. In: Blaser, MJ, Smith, PD, Ravdin, JI, Greenberg, HB, Guerrant, RL (eds) Raven Press, New York, pp 1047–1053.

Hertel, NT, Pedersen, FK, and Heilmann, C (1989) Coxsackie B3 virus encephalitis in a patient with agammaglobulinaemia. Eur J Pediatr 148:642–643.

Hierholzer, JC (1992) Adenoviruses in the immunocompromised host Clin Microbiol Rev 5:262–274.

Hirayama, M, Tokuda, A, Mutoh, T, and Kuriyama, M (1998) [Coxsackie virus B4 encephalitis in a young female who developed mental symptoms, and consciousness disturbance, and completely recovered.] Rinsho Shinkeigaku 38:60–62.

Ho, M, Chen, ER, Hsu, KH, Twu, SJ, Chen, KT, Tsai, SF, Wang, JR, and Shih, SR (1999) An epidemic of enterovirus 71 infection in Taiwan. Taiwan Enterovirus Epidemic Working Group. N Engl J Med 341:929–935.

Hollinger, FB, and Ticehurst, JR (1996) Hepatitis A virus. Virology. Lippincott-Raven, Philadelphia.

Honeymann, MC, Coulson, BS, Stone, NL, Gellert, SA, Goldwater, PN, Steele, CE, Couper, JJ, Tait, BD, Colman, PG, and Harrison, LC (2000) Association between rotavirus infection an pancreatic islet autoimmunity in children at risk of developing type 1 diabetes. Diabetes 49:1319–1324.

Hopkins, RS, Gaspard, GB, Williams, FP, Jr, Karlin, RJ, Cukor, G, and Blacklow, NR (1984) A community waterborne gastroenteritis outbreak: evidence for rotavirus as the agent. Am J Public Health 74:263–265.

Horwitz, MS (1996) Adenoviruses. Virology. Lippincott-Raven, Philadelphia.

Hurst, CJ, McClellan, KA, and Benton, WH (1988) Comparison of cytopathogenicity, immunofluorescence and in situ hybridization as methods for the detection of adenoviruses. Water Res 22:1547–1552.

Hutto, C, Little, EA, Ricks, R, Lee, JD, and Pass, RF (1986) Isolation of cytomegalovirus from toys and hands in a day-care center. J Infect Dis 154:527–530.

Hyams, KC, Purdy, MA, Kaur, M, McCarthy, MC, Hussain, MA, el-Tigani, A, Krawczynski, K, Bradley, DW, and Carl, M (1992) Acute sporadic hepatitis E in Sudanese children: analysis based on a new Western blot assay. J Infect Dis 165:1001–1005.

Hyoty, H, Hiltunen, M, and Lonnrot, M (1998) Enterovirus infections and insulin dependent diabetes mellitus – evidence for causality. Clin Diagn Virol 9:77–84.

Inouye, S, Yamashita, K, Yamadera, S, Yoshikawa, M, Kato, N, and Okabe, N (2000) Surveillance of viral gastroenteritis in Japan: Pediatric cases and outbreak incidents. J Infect Dis 181(suppl 2):S270–S274.

Irvine, DH, Irvine, AB, and Gardner, PS (1967) Outbreak of E.C.H.O. virus type 30 in a general practice. Br Med J 4:774–776.

Irving, LG, and Smith, FA (1981) One-year survey of enteroviruses, adenoviruses, and reoviruses isolated from effluent at an activated-sludge purification plant. Appl Environ Microbiol 41:51–59.

Jacobs, RF, Wilson, CB, Smith, AL, and Haas, JE (1983) Age-dependent effects of aminobutyryl muramyl dipeptide on alveolar macrophage function in infant and adult Macaca monkeys. Am Rev Respir Dis 128:862–867.

Jenista, JA, Powell, KR, and Menegus, MA (1984) Epidemiology of neonatal enterovirus infection. J Pediatr 104:685–690.

Joki-Korpela, P, and Hyypia, T (1998) Diagnosis and epidemiology of echovirus 22 infections. Clin Infect Dis 27:129–136.

Kapikian, AZ (1996) Overview of viral gastroentritis. Arch Virol 12:7–19.

Kapikian, AZ (1997) Viral Gastroenteritis. Viral Infections of Humans. Plenum, New York.

Kapikian, AZ, and Chanock, RM (1996) Rotaviruses. Virology. Lippincott-Raven, Philadelphia.

Kapikian, AZ, Wyatt, RG, Dolin, R, Thornhill, TS, Kalica, AR, and Chanock, RM (1972) Visualization by immune electron microscopy of a 27-nm particle associated with acute infectious nonbacterial gastroenteritis. J Virol 10:1075–1081.

Kaplan, JE, Gary, GW, Baron, RC, Singh, N, Schonberger, LB, Feldman, R, and Greenberg, HB (1982) Epidemiology of Norwalk gastroenteritis and the role of Norwalk virus in outbreaks of acute nonbacterial gastroenteritis. Ann Intern Med 96:756–761.

Kaplan, MH, Klein, SW, Mcphee, J, and Harper, RG (1983) Group B coxsackievirus infections in infants younger than three months of age: a serious childhood illness. Rev Infect Dis 5:1019–1032.

Karzon, DT, Eckert, GL, Barron, AL, Hayner, NS, and Winkelstein W, Jr (1961) Aseptic meningitis epidemic due to Echo 4 virus. Am J Dis Child 101:102–114.

Katz, M, and Plotkin, SA (1967) Minimal infective dose of attenuated poliovirus for man. Am J Public Health 57:1837–1840.

Kee, F, McElroy, G, Stewart, D, Coyle, P, and Watson, J (1994) A community outbreak of echovirus infection associated with an outdoor swimming pool. J Public Health Med 16:145–148.

Kennedy, JD, Talbot, IC, and Tanner, MS (1986) Severe pancreatitis and fatty liver progressing to cirrhosis associated with Coxsackie B4 infection in a three year old with alpha-1-antitrypsin deficiency. Acta Paediatr Scand 75:336–339.

Khan, NU, Gibson, A, and Foulis, AK (1990) The distribution of immunoreactive interferon-alpha in formalin-fixed paraffin-embedded normal human foetal and infant tissues. Immunology 71:230–235.

Khatib, R, Chason, JL, Silberberg, BK, and Lerner, AM (1980) Age-dependent pathogenicity of group B coxsackieviruses in Swiss-Webster mice: infectivity for myocardium and pancreas. J Infect Dis 141:394–403.

Kilgore, PE, Holman, RC, Clarke, MJ, and Glass, RI (1995) Trends of diarrheal disease–associated mortality in US children, 1968 through 1991. JAMA 274:1143–1148.

Knip, M, and Akerblom, HK (1998) IDDM prevention trials in progress – a critical assessment. J Pediatr Endocrin Metab 11(suppl 2):371–377.

Knip, M, and Akerblom, HK (1999) Environmental factors in the pathogenesis of type 1 diabetes mellitus. Exp Clin Endocrin Diabetes 107(suppl 3):S93–S100.

Koopmans, M, Vinje, J, de Wit, M, Leenen, I, van der Poel, W, and van Duynhoven, Y (2000) Molecular epidemiology of human enteric caliciviruses in the Netherlands. J Infect Dis 181(suppl 2):S262–S269.

Kosek, M, Bern, C, Caryn, RL, and Guerrant, RL (2003) The global burden of diarrhoeal disease, as estimated from studies published between 1992 and 2000. Bull WHO 81:197–204.

Krikelis, V, Markoulatos, P, Spyrou, N, and Serie, C (1985a) Detection of indigenous enteric viruses in raw sewage effluents of the city of Athens, Greece, during a two-year survey. Water Sci Technol 17:159–164.

Krikelis, V, Spyrou, N, Markoulatos, P, and Serie, C (1985b) Seasonal distribution of enteroviruses and adenoviruses in domestic sewage. Can J Microbiol 31:24–25.

Kukkula, M, Arstila, P, Klosser, ML, Maunula, L, Bonsdorff, CH, and Jaatinen, P (1997) Waterborne outbreak of viral gastroenteritis. Scand J Infect Dis 29:415–418.

Kunin, CM (1964) Cellular susceptibility to enteroviruses. Bacteriol Rev 28:382–390.

Kurtz, JB, and Lee, TW (1987) Astroviruses: human and animal. Ciba Found Symp 128:92–107.

Lane, RJ, Soteriou, BA, Zhang, H, and Archard, LC (2003) Enterovirus related metabolic myopathy: a postviral fatigue syndrome. J Neurol Neurosurg Psychiatry 74:1361–1362.

LeBaron, CW, Furutan, NP, Lew, JF, Allen, JR, Gouvea, V, Moe, C, and Monroe, SS (1990) Viral agents of gastroenteritis. Public health importance and outbreak management. Morb Mortal Wkly Rep 39(RR-5):1–24.

Leite, JP, Pereira, HG, Azeredo, RS, and Schatzmayr, HG (1985) Adenoviruses in faeces of children with acute gastroenteritis in Rio de Janeiro, Brazil. J Med Virol 15:203–209.

Lepow, ML, Warren, RJ, Ingram, VG, Daugherty, SC, Robbins, FC (1962) Sabin type I (LSc2ab) oral poliomyelitis vaccine. Effect of dose upon response on newborn infants. Am J Dis Chil 104:67–71.

Lerner, AM, Klein, JO, Cherry, JD (1963) New viral exanthems. N Engl J Med 269:678–685.

Libman, I, Songer, T, and LaPorte, R (1993) How many people in the U.S. have IDDM? Diabetes Care 16:841–842.

Lippy, EC, and Waltrip, SC (1984) Waterborne disease outbreaks – 1946–1980: A thirty-five year perspective. J Am Water Works Assoc 76:60–67.

Lonnrot, M, Korpela, K, Knip, M, Ilonen, J, Simell, O, Korhonen, S, Savola, K, Muona, K, Simell, T, Koskela, P, and Hyoty, H (2000) Enterovirus infection as a risk factor for beta-cell autoimmunity in a prospectively observed birth cohort: the Finnish Diabetes Prediction and Prevention Study. Diabetes 49:1314–1318.

Loria, RM, Shadoff, N, Kibrick, S, and Broitman, S (1976) Maturation of intestinal defenses against peroral infection with group B coxsackievirus in mice. Infect Immun 13:1397–1401.

Lucioni, C, Cipriani, V, Mazzi, S, and Panunzio, M (1998) Cost of an outbreak of hepatitis A in Puglia, Italy. Pharmacoeconomics 13:257–266.

Madeley, CR, and Cosgrove, BP (1975) Letter: 28-nm particles in faeces in infantile gastroenteritis. Lancet 2:451–452.

Margolis, HS, Alter, MJ, and Hadler, SC (1997) Viral Hepatitis. Viral Infections of Humans. Plenum, New York.

Marlin, AE, Huntington, WH, Arizpe, HM, Gudvangen, RJ, Brans, YW, and Gauntt, CJ (1985) Coxsackie group B and hydronencephaly. Concepts Pediatr Neurosurg 6:147–160.

Martino, TA, Liu, P, Petric, M, and Sole, MJ (1995) Enteroviral myocarditis and dilated cardiomyopathy: A reveiw of clinical and experimental studies. In: Rotbart, HA (ed) Human Enterovirus Infections. ASM Press, Washington, DC, pp. 291–351.

Martinson, FE, Marfo, VY, and Degraaf, J (1999) Hepatitis E virus seroprevalence in children living in rural Ghana. West Afr J Med 18:76–79.

Martone, WJ, Hierholzer, JC, Keenlyside, RA, Fraser, DW, D'Angelo, LJ, and Winkler, WG (1980) An outbreak of adenovirus type 3 disease at a private recreation center swimming pool. Am J Epidemiol 111:229–237.

McGee, ZA, and Baringer, JR (1990) Acute meningitis. In: Mandell, GL, Douglas, RG, Bennett, JE (eds) Principles and Practice of Infectious Diseases. Churchill Livingstone, New York, pp. 741–755.

McKinney, RE, Kotz, SL, and Wilfert, CM (1987) Chronic enteroviral meningoencephalitis in agammaglobulinemic patients. Rev Infect Dis 9:334–356.

McMillian, NS, Martin, SA, Sobsey, MD, Wait, DA, Meriwether, RA, and MacCormack, JN (1992) Outbreak of pharnygoconjunctival fever in a summer camp – North Carolina, 1991. Morb Mortal Wkly Rep 41:342–370.

McMinn, PC, Stewart, J, Burrell JC (1991) A community outbreak of epidemic keratoconjunctivitis in central Australia due to adenovirus type 8. J Infect Dis 164:1113–1118.

Melnick, JL (1984) Enterovirus type 71 infections: a varied clinical pattern sometimes mimicking paralytic poliomyelitis. Rev Infect Dis 6(suppl 2):S387–S390.

Melnick, JL (1997) Poliovirus and other enteroviruses. In: Evans, AS, Kaslow, RA (eds) Viral Infections of Humans. Plenum, New York, pp. 583–663.

Meng, XJ, Halbur, PG, Shapiro, MS, Govindarajan, S, Bruna, JD, Mushahwar, IK, Purcell, RH, and Melnick Emerson, SU (1998) Genetic and experimental evidence for cross-species infection by swine hepatitis E virus. J Virol 72:9714–9721.

Meng, XJ, Dea, S, Engle, RE, Friendship, R, Lyoo, YS, Sirinarumitr, T, Urairong, K, Wang, D, Wwong, D, Zhang, Y, Prucell, RH, and Emerson, SU (1999) Prevalence of antibodies to the hepatitis E virus in pigs from countries where hepatitis E is common or is rare in the human population. J Med Virol 59:297–302.

Minor, TE, Allen, CI, Tsiatis, AA, Nelson, DB, and D'Alessio, DJ (1981) Human infective dose determinations for oral poliovirus type 1 vaccine in infants. J Clin Microbiol 13:388–389.

Mitchell, DK, Matson, DO, Cubitt, WD, Jackson, LJ, Willcocks, MM, Pickering, LK, and Carter, MJ (1999) Prevalence of antibodies to astrovirus types 1 and 3 in children and adolescents in Norfolk, Virginia. Pediatr Infect Dis J 18:249–254.

Modlin, JF (1986) Perinatal echovirus infection: insights from a literature review of 61 cases of serious infection and 16 outbreaks in nurseries. Rev Infect Dis 8:918–926.

Modlin, JF (1990) Coxsackieviruses, echoviruses, and newer enteroviruses. In: Mandell, GL, Douglas, RG, Bennett, JE (eds) Principles and Practice of Infectious Diseases. Churchill Livingstone, New York. pp. 1367–1387.

Modlin, JF, and Rotbart, HA (1997) Group B coxsackie disease in children. Curr Topics Microbiol Immunol 223:53–80.

Mohle-Boetani, JC, Matkin, C, Pallansch, M, Helfand, R, Fenstersheib, M, Blanding, JA, and Solomon, SL (1999) Viral meningitis in child care center staff and parents: an outbreak of echovirus 30 infections. Public Health Rep 114:249–256.

Moore, M (1992) Enteroviral disease in the United States. J Infect Dis 146:103–108.

Moreira, RC, Castrignano, SB, Carmona, RDC, Gomes, FM, Saes, SG, Oliveira, RS, Souza, DF, Takimoto, S, Costa, MC, and Waldman, EA (1995) An exanthematic disease epidemic associated with coxsackievirus B3 infection in a day care center. Rev Instit Med Trop Sao Paulo 37:235–238.

Morens, DM, Pallansch, MA, and Moore, M. (1991) Polioviruses and other enteroviruses. In: Belsche, RB (ed) Textbook of Human Virology. Mosby Year Book, New York, pp. 425–497.

Nairn, C, Galbraith, DN, Taylor, KW, and Clements, GB (1999) Enterovirus variants in the serum of children at the onset of type 1 diabetes mellitus. Diabetic Med 16:509–513.

NAS (1983) Assessment in the Federal Government: Managing the Process. National Academy Press, Washington, DC.

Newman, RD, Grupp-Phelan, J, Shay, DK, and Davis, RL (1999) Perinatal risk factors for infant hospitalizations with viral gastroenteritis. Pediatrics 103:e3.

Numata, K, Nakata, S, Jiang, X, Estes MK, Chiba, S (1994) Epidemiological study of Norwalk virus infections in Japan and Southeast Asia by enzyme-linked immunosorbant assays with Norwalk virus Capsid protein produced by the baculovirus expression system. J Clin Microbiol 32:121–126.

O'Connor, JB, Imperiale, TF, and Singer, ME (1999) Cost-effectiveness analysis of hepatitis A vaccination strategies for adults. Hepatology 30:1077–1081.

O'Ryan, ML, Matson, DO, Estes, MK, Bartlett, AV, and Pickering, LK (1990) Molecular epidemiology of rotavirus in children attending day care centers in Houston. J Infect Dis 162:810–816.

Ostroff, SM, and Leduc, JW (2000) Global epidemiology of infectious diseases. In: Mandell, GL, Bennett, JE, Dolin, R. (ed) Principles and Practices of Infectious Diseases. Churchill Livingstone, Philadelphia, pp. 167–178.

Pallansch, MA (1997) Coxsackievirus B epidemiology and public health concerns. Curr Topics Microbiol Immunol 223:13–30.

Pang, DT, Phillips, CL, and Bawden, JW (1992) Fluoride intake from beverage consumption in a sample of North Carolina children. J Dent Res 71:1382–1388.

Papapetropoulou, M, and Vantarakis, AC (1998) Detection of adenovirus outbreak at a municipal swimming pool by nested PCR amplification. J Infect 36:101–103.

Parashar, UD, Holman, RC, Clarke, MJ, Bresee, JS, and Glass, RI (1998) Hospitalizations associated with rotavirus diarrhea in the United States, 1993 through 1995: surveillance based on the new ICD-9-CM rotavirus-specific diagnostic code. J Infect Dis 177:13–17.

Parashar, UD, Chung, MA, Holman, RC, Ryder, RW, Hadler, JL, and Glass, RI (1999) Use of state hospital discharge data to assess the morbidity from rotavirus diarrhea and to monitor the impact of a rotavirus immunization program: A pilot study in Connecticut. Pediatrics 104:489–494.

Parashar, UD, Hummelman, EG, Bresee, JS, Miller, MA, and Glass, RI (2003) Global illnesses and deaths caused by rotavirus disease in children. Emerg Infect Dis 9:565–572.

Payment, P, Tremblay, M, and Trudel, M (1985) Relative resistance to chlorine of poliovirus and coxsackievirus isolates from environmental sources and drinking water. Appl Environ Microbiol 49:981–983.

Payment, P, Richardson, L, Siemiatycki, J, Dewar, R, Edwardes, M, and Franco, E (1991) A randomized trial to evaluate the risk of gastrointestinal disease due to consumption of drinking water meeting current microbiological standards. Am J Public Health 81:703–708.

Payment, P, Siemiatycki, J, Richardson, L, Renaud, G, Franco, E, and Prevost, M (1997) A prospective epidemiology study of gastrointestinal health effects due to the consumption of drinking water. Int J Environ Health Res 7:5–31.

Perera, BJC, Ganesan, S, Jayarasa, J, and Ranaweera, S (1999) The impact of breast-feeding practices on respiratory and diarrhoeal disease in infancy: a study from Sri Lanka. J Trop Pediatr 45:115–118.

Peterson, ML (1974) Soiled disposable diapers: A potential source of viruses. Am J Public Health 64:912–914.

Petric, M, Krajden, S, Dowbnia, N, and Middleton, PJ (1982) Enteric adenoviruses [letter]. Lancet 1:1074–1075.

Pichichero, ME, McLinn, S, Rotbart, HA, Menegus, MA, Cascino, M, and Reidenberg, BE (1998) Clinical and economic impact of enterovirus illness in private pediatric practice. Pediatrics 102:1126–1134.

Pickering, LK, Granoff, DM, Erickson, JR, Masor, ML, Cordle, CT, Schaller, JP, Winship, TR, Paule, CL, and Hilty, MD (1998) Modulation of the immune system by human milk and infant formula containing nucleotides. Pediatrics 101:242–249.

Plotkin, A, and Katz, M (1967) Mimimal infective doses of viruses for man by the oral route. In: Berg, G (ed) Transmission of Viruses by the Water Route. Wiley Inter-science, New York, pp. 151–166.

Plotkin, SA, Koprowski, H, Stokes J Jr (1959) Clinical trails infants of rally adminis-trated attenuated poliomyelitis viruses. Pediatrics 23:1041–1102.

Puig, M, Jofre, J, Lucena, F, Allard, A, Wadell, G, and Girones, R (1994) Detection of adenoviruses and enteroviruses in polluted waters by nested PCR amplifica-tion. Appl Environ Microbiol 60:2963–2970.

Ramachandran, M, Gentsch, JR, Parashar, UD, Jin, S, Woods, PA, Holmes, JL, Kirkwood, CD, Bishop, RF, Greenberg, HB, Urasawa, S, Gerna, G, Coulson, BS, Taniguchi, K, Bresee, JS, and Glass, RI (1998) Detection and characteriza-tion of novel rotavirus strains in the United States. J Clin Microbiol 36:3223–3229.

Rao, VC, and Melnick, JL (1986) Environmental Virology. American Society for Microbiology. Washington DC.

Rantakallio, P, Jones, P, Moring, J, and Von Wendt, L (1997) Association between central nervous system infections during childhood and adult onset schizophre-nia and other psychoses: a 28-year follow-up. Int J Epidemiol 26:837–843.

Regli, S, Rose, JB, Haas, CN, and Gerba, CP (1991) Modeling the risk from *Giardia* and viruses in drinking water. J Am Water Works Assoc 83:76–84.

Reina, J, Hervas, J, and Ros, MJ (1994) [Differential clinical characteristics among pediatric patients with gastroenteritis caused by rotavirus and adenovirus.] Enferm Infecc Microbiol Clin 12:378–384.

Rewers, M, and Atkinson, M (1995) The possible role of enteroviruses in diabetes mellitis. In: Rotbart, HA (ed) Human Enterovirus Infections. ASM Press, Washington, DC, pp 353–385.

Rice, SK, Heinl, RE, Thornton, LL, and Opal, SM (1995) Clinical characteristics, management strategies, and cost implications of a statewide outbreak of enterovirus meningitis. Clin Infect Dis 20:931–937.

Rodriguez, WJ, Kim, HW, Brandt, CD, Schwartz, RH, Gardner, MK, Jeffries, B, Parrott, RH, Kaslow, RA, Smith, JI, and Kapikian, AZ (1987) Longitudinal study of rotavirus infection and gastroenteritis in families served by a pediatric medical practice: clinical and epidemiologic observations. Pediatr Infect Dis J 6:170–176.

Roper, WL, Murphy, FA, Mahy, BWJ, Anderson, LS, and Glass, RI (1990) Viral agents of gastroenteritis: Public health importance and outbreak management. Morb Mortal Wkly Rep 39:1–24.

Rorabaugh, ML, Berlin, LE, Heldrich, F, Roberts, K, Rosenberg, LA, Doran, T, and Modlin, JF (1993) Aseptic meningitis in infants younger than 2 years of age: acute illness and neurologic complications. Pediatrics 92:206–211.

Rose, JB, Haas, CN, and Gerba, CP (1996) Risk assessment for microbial contaminants in water. Report for the AWWA Research Foundation, Denver, CO.

Roseberry, AM, and Burmaster, DE (1992) Lognormal distributions for water intake by children and adults. Risk Anal 12:99–104.

Roy, CC, Silverman, A (eds) (1995) Pediatric Clinical Gastroenterology, 3rd Ed. Mosby, St. Louis.

Ryan, MJ, Ramsay, M, Brown, D, Gay, NJ, Farrington, CP, and Wall, PG (1996) Hospital admissions attributable to rotavirus infection in England and Wales. J Infect Dis 174(suppl 1):S12–S18.

San Pedro, MC, and Waltz, SE (1991) A comprehensive survey of pediatric diarrhea at a private hospital in metro Manila. Southeast Asian J Trop Med Public Health 22:203–210.

Sedman, RM, and Mahmood, RJ (1994) Soil ingestion by children and adults reconsidered using the results of recent tracer studies. J Air Waste Manag Assoc 44:141–144.

Shattuck, KE, and Chonmaitree, T (1992) The changing spectrum of neonatal meningitis over a fifteen-year period. Clin Pediatr 31:130–136.

Shaw, FE, Jr, Sudman, JH, Smith, SM, Williams, DL, Kapell, LA, Hadler, SC, Halpin, TJ, and Maynard, JE (1986) A community-wide epidemic of hepatitis A in Ohio. Am J Epidemiol 123:1057–1065.

Sherman, PM, and Litchman, SN (1995) Pediatric considerations relevant to enteric infections. In: Blaser, MJ, Ravdin, JI, Smith, PD, Greenberg, HB, and Guerrant, RL (eds) Infections of the Gastrointestinal Tract. Raven Press, New York, pp. 143–152.

Simonsen, L, Khan, AS, Gary, HE, Jr, Hanson, C, Pallansch, MA, Music, S, Holman, RC, Stewart, JA, Erdman, DD, Arden, NH, Arenberg, IK, and Schonberger, LB (1996) Outbreak of vertigo in Wyoming: possible role of an enterovirus infection. Epidemiol Infect 117:149–157.

Smith, CP, Clements, GB, Riding, MH, Collins, P, Bottazzo, GF, and Taylor, KW (1998) Simultaneous onset of type 1 diabetes mellitus in identical infant twins with enterovirus infection. Diabet Med 15:515–517.

Smith, JC, Haddix, AC, Teutsch, SM, and Glass, RI (1995) Cost-effectiveness analysis of a rotavirus immunization program for the United States. Pediatrics 96 (4 pt 1):609–615.

Spratt, HC, Marks, MI, Gomersall, M, Gill, P, and Pai, CH (1978) Nosocomial infantile gastroenteritis associated with minirotavirus and calicivirus. J Pediatr 93:922–926.

Springthorpe, VS, and Sattar, SA (1990) Chemical disinfection of virus contaminated surfaces. CRC Crit Rev Environ Control 20:169–229.

Steele, AD, and Sears, JF (1996) Characterization of rotaviruses recovered from neonates with symptomatic infection. S Afr Med J 86:1546–1549.

Steer, RG (1992) Echovirus 16 orchitis and postviral fatigue syndrome. Med J Aust 156:816.

Stewien, KE, da Cunha, LC, Alvim, ADC, dos Reis Filho, SA, Alvim, MA, Brandao, AA, and Neiva, MN (1991) Rotavirus associated diarrhoea during infancy in the city of S. Luis (MA), Brazil: a two-year longitudinal study. Rev Instit Med Trop Sao Paulo 33:459–464.

Straus, SE (1984) Adenovirus infection in humans. In: Ginsberg, HS (ed) The Adenoviruses. Plenum Press, New York. pp. 321–332.

Sutmoller, F, Azeredo, RS, Lacerda, MD, Barth, OM, Pereira, HG, Hoffer, E, and Schatzmayr, HG (1982) An outbreak of gastroenteritis caused by both rotavirus and Shigella sonnei in a private school in Rio de Janeiro. J Hyg 88:285–293.

Talusan-Soriano, K, and Lake, AM (1996) Malabsorption in childhood. Pediatr Rev 17:135–142.

Tam, JS, Kum, WW, Lam, B, Yeung, CY, and Ng, MH (1986) Molecular epidemiology of human rotavirus infection in children in Hong Kong. J Clin Microbiol 23:660–664.

Tan, CR, Chen, M, Ge, SX, Zhang, J, Hu, M, Sun, HY, Chen, Y, Peng, G, Shu, W, Zhang, M, and Xia, NS (2003) Serological characteristics of a heatitis E outbreak. Zhonghua Shi Yan He Lin Chuang Bing Du Xue Za Zhi 17:361–364.

Taylor, JW, Gary, GW, Jr, and Greenberg, HB (1981) Norwalk-related viral gastroenteritis due to contaminated drinking water. Am J Epidemiol 114:584–592.

Tei, S, Kitajima, N, Ohara S, Inoue, Y, Miki, M, Yamatani, T, Yamabe, H, Mishiro, S, and Kinoshita, Y (2004) Consumption of uncooked deer meat as a risk factor for hepatitis E virus infection: an age- and sex-matched case-control study. J Med Virol 74:67–70.

Teunis, PFM, van der Heijden, OG, van der Giessen, JWB, and Havelaar, AH (1996) The dose-response relation in human volunteers for gastrointestinal pathogens. Rep. No. 284550002. National Institute of Public Health and the Environment, Bilthoven, The Netherlands.

Thorner, PA, Ahrel-Andersson, M, Hierholzer, JC, and Johansson, ME (1993) Characterization of two divergent adenovirus 31 strains. Arch Virol 133:397–405.

Timens, W, Boes, A, Rozeboom-Uiterwijk, T, and Poppema, S (1989) Immaturity of the human splenic marginal zone in infancy. Possible contribution to the deficient infant immune response. J Immunol 143:3200–3206.

Tompkins, DS, Hudson, MJ, Smith, HR, Eglin, RP, Wheeler, JG, Brett, MM, Owen, RJ, Brazier, JS, Cumberland, P, King, V, and Cook, PE (1999) A study of infectious intestinal disease in England: microbiological findings in cases and controls [published erratum appears in Commun Dis Public Health 1999;2(3):222]. Commun Dis Public Health 2:108–113.

Uhnoo, I, Wadell, G, Svensson, L, Olding-Stenkvist, E, Ekwall, E, and Molby, R (1986) Aetiology and epidemiology of acute gastro-enteritis in Swedish children. J Infect 13:73–89.

Utagawa, ET, Nishizawa, S, Sekine, S, Hayashi, Y, Ishihara, Y, Oishi, I, Iwasaki, A, Yamashita, I, Miyamura, K, and Yamazaki, S (1994) Astrovirus as a cause of gastroenteritis in Japan. J Clin Microbiol 32:1841–1845.

Van, R, Wun, CC, O'Ryan, ML, Matson, DO, Jackson, L, and Pickering, LK (1992) Outbreaks of human enteric adenovirus types 40 and 41 in Houston day care centers. J Pediatr 120:516–521.

Velazquez, FR, Matson, DO, Calva, JJ, Guerrero, L, Morrow, AL, Carter-Campbell, S, Glass, RI, Estes, MK, Pickering, LK, and Ruiz-Palacios, GM (1996) Rotavirus infections in infants as protection against subsequent infections. N Engl J Med 335:1022–1028.

Wang, SM, Liu, CC, Chen, YJ, Chang, YC, and Huang, CC (1996) Alice in Wonderland syndrome caused by coxsackievirus B1. Pediatr Infect Dis J 15:470–471.

Welliver, RC, and Cherry, JD (1978) Aseptic meningitis and orchitis associated with echovirus 6 infection. J Pediatr 92:239–240.

Whatley, TR, Comstock, GW, Garber, HJ, and Sanchez, FS, Jr (1968) A waterborne outbreak of infectious hepatitis in a small Maryland town. Am J Epidemiol 87:138–147.

White, DO, and Fenner, FJ (1994) Medical Virology. Academic Press, San Diego.

Willcocks, MM, Brown, TD, Madeley, CR, and Carter, MJ (1994) The complete sequence of a human astrovirus. J Gen Virol 75:1785–1788.

Williams, FP, and Hurst, CJ (1988) Detection of environmental viruses in sludge: enhancement of enterovirus plaque assay titers with 5-iodo-2'-deoxyuridine and comparison to adenovirus and coliphage titers. Water Res 22:847–851.

Wilson, CB (1986) Immunologic basis for increased susceptibility of the neonate to infection. J Pediatr 108:1–12.

Woodruff, JF (1980) Viral myocarditis: A review. Am J Pathol 101:427–483.

Zaher, SR, Kassem, AS, and Hughes, JJ (1993) Coxsackie virus infections in rheumatic fever. Indian J Pediatr 60:289–298.

Zaoutis, T, and Klein, JD (1998) Enterovirus infections. Pediatri Rev 19:183–191.

Manuscript received February 1; accepted February 7, 2005.

Rev Environ Contam Toxicol 186:57–72   © Springer 2006

Pyrethroid Illnesses in California, 1996–2002

Janet Spencer and Michael O'Malley

Contents

I. Introduction

With regulatory limitations on the use of cholinesterase-inhibiting insecticides, synthetic pyrethroids have become increasingly important in both agricultural and structural pest control (USEPA 1996, 2000). Pyrethroid chemistry and action are classified as type I or type II depending on the alcohol substituent. Type I pyrethroids contain a descyano-3-phenoxybenzyl or other alcohols. The type II, or alpha-cyano pyrethroids, contain α-cyano-3-phenoxybenzyl alcohol, which increases insecticidal activity about 10-fold.

The California Department of Pesticide Regulation (DPR), Worker Health and Safety Branch (WH&S) is responsible for public and worker safety wherever pesticides are used in the state. Their mission is accomplished through several programs: exposure monitoring, exposure assessment and mitigation, pesticide illness surveillance, and workplace evaluation and industrial hygiene. The WH&S Pesticide Illness Surveillance Program (PISP) is generally acknowledged as the nation's most compre-

Communicated by George W. Ware

J. Spencer (✉) · M. O'Malley
California Department of Pesticide Regulation, Worker Health and Safety Branch, 1001 I St., P.O. Box 4015, Sacramento, CA 95812-4015 U.S.A.

hensive and reliable reporting system (Mehler et al. 1992; U.S. Government Accounting Office 1994). Data generated from illness reports have made DPR's worker protection program a model for other states (Western Farm Press 2001). WH&S has previously analyzed PISP data to develop retrospectives of illnesses, injuries, and deaths related to pesticide exposures in California to determine risk factors for exposure to organophosphate insecticides in agricultural workers and to summarize investigations of exposure incidents related to specific pesticides (Maddy et al. 1990; O'Malley et al. 1990; O'Malley 1998b; O'Malley and Verder-Carlos 2001). WH&S has also evaluated PISP data to determine the effectiveness of California's worker protection program regarding field posting, hazard communication, notification and retaliation requirements, and irrigator and pesticide handler exposures (Fong 2001; McCarthy 2003; Spencer 2001).

This survey summarizes California's recent experience with pyrethroid-related illnesses, using PISP data, pesticide use reporting data, and investigations of three large group illness episodes related to exposure to type II pyrethroids cyfluthrin and λ-cyhalothrin (Edmiston et al. 1998, 1999). Cases were reviewed by use, structural class and routes of exposure for the years 1996–2002, focusing on 317 cases involving exposure to one or more pyrethroid compounds (DPR, unpublished data).

Illnesses associated with pyrethroid exposures during packaging, during agricultural and indoor/structural applications, and resulting from indoor residue exposures have been documented (He et al. 1988; Muller-Mohnssen 1999; Pauluhn 1996; Prohl et al. 1997). Few studies have evaluated agricultural occupational exposures to pyrethroid residues (Kolmodin-Hedman et al. 1982, 1995). Although pyrethroid residues on crops, agricultural products, and foliage have previously been investigated, pyrethroid dissipation is not well understood, as compared to the dissipation of organophosphorus pesticides (Argauer et al. 1997; Bellows et al. 1993; Dejonckheere et al. 1982; Edmiston et al. 1999; Estesen and Buck 1990; Giles et al. 1992; Hernandez et al. 1998; McEwen et al. 1986; Miyata et al. 1993; Nakamura et al. 1993; Papadopoulou-Mourkidou et al. 1989). This review provides recent data on agricultural and nonagricultural illnesses related to pyrethroid exposures and augments the data available on pyrethroid residue dissipation.

II. Pyrethroid Mode of Action

Pyrethroids cause prolongation of sodium channel currents in the nervous system, and many of their adverse effects are related to this property. For all pyrethroids, the major neurotoxic hazard is acute excitation. Exposure to pyrethroids can cause dermal irritation and paresthesia, very often without producing visible erythema of the skin (Cagen et al. 1984). Similar irritant effects occur in the respiratory tract (Pauluhn et al. 1996; Pauluhn 1999; Ray 2000). Systemic toxicity has been reported on ingestion

(O'Malley 1997) and may also occur with a high degree of occupational exposure (He et al. 1988, 1989). The α-cyano subgroup of compounds (type II pyrethroids) produce more prolonged sodium channel currents than type I compounds and often are relatively more potent as insecticides and mammalian toxins (Casida et al. 1983). In laboratory animals, the toxidromes produced by type I and type II compounds are considered distinct (Ray 2000). Type I pyrethroids cause the simplest poisoning syndrome, characterized in animal studies by severe fine tremor, marked reflex hyperexcitability, sympathetic activation, and, for dermal exposure, paresthesia. Type II pyrethroid poisoning is more complex and causes more severe symptoms. In addition to sympathetic activation and paresthesia, the toxidrome exhibited profuse watery salivation, coarse tremor, increased extensor tone, moderate reflex hyperexcitability, choreoathetosis, and seizures (Ray 2000).

III. Illness, Exposure, and Pesticide Use Data
A. Illness and Use Report Data

California has implemented a full-use reporting system for pesticides since 1990. Since 1981, DPR's PISP has maintained a database of pesticide-related illnesses and injuries. DPR receives case reports from physicians and via Workers' Compensation records. The local county agricultural commissioner investigates circumstances of exposure for each case. WH&S then evaluates the medical records and investigative findings and enters the data into an illness registry. WH&S may conduct investigations when it appears that significant hazards contributed to the exposures, such as in the three group episodes of exposure to type II pyrethroids discussed later.

Table 1 summarizes both use and illness data associated with the 317 pyrethroid-related illnesses reported to PISP between 1996 and 2002. For the 13 active ingredients associated with those illnesses, 4,629,851 lb (2,100,068 kg) were reported to DPR as being used in the state during the same time period (Table 1) (DPR 2004). Two compounds, permethrin and cypermethrin, accounted for 75% of total use. Type II compounds accounted for 42.8% of the reported pounds used (1,979,352 lb; 897,820 kg), but were associated with 220 (69.4%) of the cases. A single compound, cyfluthrin, was associated with 122 cases (55% of illnesses related to type II pyrethroids and 38.4% of all pyrethroid illnesses). Cyfluthrin use (308,191 lb; 139,793 kg) comprised 15.6% of total type II usage and 6.6% of total pyrethroid usage. Type I compounds accounted for 2,650,500 lb (1,202,248 kg), 57.2% of the reported use, and 97 (30.6%) of the reported illness cases. The type I compounds most frequently associated with illnesses were resmethrin (38 cases) and permethrin (44 cases).

Agricultural use includes uses for treatment of crops, nurseries, livestock, agricultural research facilities, and the handling of raw agricultural commodities in packing houses. These uses accounted for 118 (37.3%) of the

Table 1. Summary of Pyrethroid-Related Illnesses[a] and Reported Pyrethroid Use[b] in California, 1996–2002.

Pyrethroid	No. illnesses	Reported use (lb)	Reported use (kg)
Type I			
Allethrin	3	571	259
Bifenthrin	11	199,302	90,402
Permethrin	44	2,445,525	1,109,273
Phenothrin	1	583	264
Resmethrin	38	4,519	2,050
Total Type I	97	2,650,500	1,202,248
Type II			
Cyfluthrin	122	308,191	139,793
Cypermethrin	43	1,091,419	495,060
Deltamethrin	3	45,216	20,510
Esfenvalerate	26	234,886	106,543
Fenpropathrin	2	117,637	53,359
Fluvalinate	3	19,929	9,040
λ-Cyhalothrin	12	153,621	69,681
Tralomethrin	9	8,453	3,834
Total Type II	220	1,979,352	897,820
Total Use		4,629,852	2,100,068

[a]Reported to California's Pesticide Illness Surveillance Program; illness was related to one or more pyrethroids as the primary causal agent.

[b]Reporting required for agricultural use, all use by pest control businesses, all restricted materials, all institutional use of pesticides on the groundwater protection list, and all post-harvest commodity treatment; does not include home and garden use of products sold over-the-counter.

reported cases, with all but 2 cases (related to bystander drift) associated with employment. Agricultural cases were primarily (95%) associated with applications to either crops (86 illnesses, 62 of which were in oranges) or food processing/storage facilities (26 illnesses). For the 199 cases (62.8%) associated with nonagricultural use, 132 (66.3%) were occupationally related. Nonagricultural cases were predominantly (80%) related to applications inside or outside of buildings and homes (159 illnesses). Nonagricultural uses included residential, retail, service and institutional uses, structural pest control, rights-of-way, parks, landscaped urban areas, and public health vector control (Table 2).

B. Group Versus Individual Illness

Approximately equal numbers of illnesses resulted from individual exposures (167 cases) and group exposures (150 cases). The majority of illnesses related to agricultural use (75%) occurred in seven group episodes (88 ill-

Table 2. Summary of Pyrethroid-Related Illnesses Reported to California's Pesticide Illness Surveillance Program, 1996–2002, by Number of Persons Involved in Group Illness Episodes[a] versus Individual Illnesses.[b]

| Year | No. persons exposed in group illness episodes[a] (no. episodes) | | No. persons exposed in individual illnesses[b] | | Total |
	Agricultural[c]	Nonagricultural[d]	Agricultural[c]	Nonagricultural[d]	
1996	2 (1)	6 (1)	1	16	25 (2)
1997	55 (2)	10 (5)	9	22	96 (7)
1998	0	7 (1)	4	7	18 (1)
1999	0	21 (3)	4	15	40 (3)
2000	26 (2)	9 (4)	4	25	64 (6)
2001	5 (2)	9 (2)	4	17	35 (4)
2002	0	0	4	35	39 (0)
Total	88 (7)	62 (16)	30	137	317 (23)

[a]Episode: a single exposure event; one or more persons may became ill from a single pesticide exposure incident.
[b]Illness: indicates a single individual became ill.
[c]Uses intended to contribute to the production of agricultural commodities including crops, nursery products, and livestock. It includes transportation and storage of pesticides on farmlands and use at agricultural research facilities and in packing houses. It excludes forestry operations and the manufacture, transportation, and storage of pesticides before arrival at the site of agricultural production. Pesticides used agriculturally retain their agricultural designation regardless of exposure location.
[d]The pesticide(s) were not intended to contribute to the production of agricultural commodities. This includes residential, retail, service, and institutional uses, structural pest control, use in rights-of-way, parks, and landscaped urban areas, and the manufacture, transportation, and storage of pesticides except on farmlands.

Table 3. Summary of Symptoms for Pyrethroid-Related Illnesses Reported to California's Pesticide Illness Surveillance Program, 1996–2002.

Symptom array	Type I pyrethroids	Type II pyrethroids	Total illnesses by symptom array
Irritant[a]	26	107	133
Irritant/systemic	41	95	136
Systemic	30	18	48
Total illnesses	97	220	317

[a]Dermal, eye, and/or respiratory irritant symptoms.

nesses). The majority of nonagricultural use-related illnesses (69%) were due to individual exposures (137 illnesses).

C. Symptom Array and Exposure Route

The symptom arrays associated with the pyrethroid illnesses are summarized in Table 3. Irritant effects or paresthesias of the eye, skin, or respiratory tract were present in 269 cases (84.9%). Type II pyrethroids were more frequently associated with isolated irritant symptoms (107 cases) than the type I pyrethroids (26 of 97 cases). Systemic symptoms were reported in 184 illnesses (58% of cases). Isolated systemic effects occurred in 48 cases (15.1%), but systemic effects were also present in 136 (50.6%) of the 269 cases with irritant symptoms. Isolated systemic symptoms were reported in 30 of 97 cases related to type I pyrethroids and in 18 of 220 cases involving type II pyrethroids. Specific symptoms reported included headache, nausea, vomiting, epigastric pain, weakness, lethargy, fatigue, dizziness, sweating, and/or muscle pain.

Residue exposures accounted for 158 illnesses (49.8%), with the majority of the illnesses (149, 94%) occurring in the following activity categories: field worker (69 cases), routine indoor exposures (55 cases), and packing/processing (25 cases) (Table 4). Direct exposures accounted for 75 illnesses (23.7%), with the majority occurring while persons were involved in application (39 cases) and mixing/loading (9 cases) tasks, emergency response activities (8 cases), and activities categorized as other (8 cases). Drift exposures accounted for 62 illnesses (19.6%), with the majority occurring during routine indoor activities (34 cases), application tasks (15 cases), and routine outdoor activities (9 cases). Because of the three illness clusters, the residue exposure cases accounted for a large majority (79.7%) of the 118 agricultural cases. Only 32.1% of the 199 nonagricultural cases were associated with residue exposure. Seven illnesses resulted from intentional or accidental ingestion and 15 had unknown/other exposure routes. The latter group included notable illnesses or injuries following explosions that occurred when excessive application of indoor foggers ignited vapors from pilot lights.

D. Pesticide Use Violations

Single or multiple violations of pesticide use regulations contributed to exposures in 90 of the 317 illnesses (28.4%); 76 were related to nonagricultural pyrethroid use. The most common violations involved pesticide misuse (78 cases), including failure to control drift, failure to notify building occupants of a structural application, and use of higher than label rates in applications (including residents using multiple aerosol foggers to treat dwellings). Other contributing violations, alone or in combination with pesticide misuse, included failure to wear proper protective equipment and early reentry following an application.

IV. Three Group Illness Episode Investigations

The following summarizes WH&S investigations of three group illness episodes where respiratory and dermal irritation symptoms were prominent. The episodes were related to agricultural exposures to type II

Table 4. Summary of Pyrethroid-Related Illnesses Reported to California's Pesticide Illness Surveillance Program, 1996–2002, by Symptom Array, Exposure Route, and the Relationship to Agricultural[a] or Nonagricultural[b] Pesticide Use.

Exposure route	Total	Symptom array		
		Irritant[c]	Irritant/systemic	Systemic
Direct[d]	75	45	18	12
Agricultural[a]	11	10	0	1
Nonagricultural[b]	64	35	18	11
Drift[e]	62	20	30	12
Agricultural	12	4	4	4
Nonagricultural	50	16	26	8
Residue[f]	158	57	81	20
Agricultural	94	31	50	13
Nonagricultural	64	26	31	7
Ingestion[g]	7	0	4	3
Unknown/other	15	11	3	1
Agricultural	1	0	0	1
Nonagricultural	14	11	3	0
Grand totals	317	133	136	48

[a]Uses intended to contribute to the production of agricultural commodities including crops, nursery products, and livestock. It includes transportation and storage of pesticides on farmlands and use at agricultural research facilities and in packing houses. It excludes forestry operations and the manufacture, transportation, and storage of pesticides before arrival at the site of agricultural production. Pesticides used agriculturally retain their agricultural designation regardless of exposure location.
[b]The pesticide(s) were not intended to contribute to the production of agricultural commodities. This includes residential, retail, service, and institutional uses, structural pest control, use in rights-of-way, parks, and landscaped urban areas, and the manufacture, transportation, and storage of pesticides except on farmlands.
[c]Dermal, eye, and/or respiratory irritant symptoms.
[d]Body contact with appreciable quantities of pesticide.
[e]Airborne exposure during pesticide application or preparation for application.
[f]Exposure to any amount or component of a pesticide that remains in the environment after application.
[g]Accidental or intentional ingestions.

pyrethroids cyfluthrin (two episodes) and λ-cyhalothrin (one episode) (DPR, unpublished data; Edmiston et al. 1998, 1999). The cyfluthrin illnesses were included among the 317 discussed previously. The λ-cyhalothrin illnesses were not included in the preceding discussion because propargite and sulfur residues were also present. Although all three episodes involved harvesters entering treated fields well after applicable restricted entry intervals had elapsed, the product in the λ-cyhalothrin episode had been mistakenly and illegally applied to grapes, at rates far higher than legally permitted for registered crop uses.

A. Cyfluthrin Orange Harvester Episodes

Cyfluthrin was involved in 121 of the 317 illnesses described previously. Sixty-nine illnesses were related to agricultural uses, with 61 illnesses occurring in oranges: the two group episodes described below involved 55 harvesters. The other 6 illnesses included three field workers exposed in individual illnesses, two applicators and one mixer/loader. Fifty-two illnesses occurred in nonagricultural use settings, predominantly (38 illnesses) affecting persons exposed to residue (29 illnesses) or drift (9 illnesses) while involved in routine indoor activities.

The two group illness episodes occurred in Tulare County in May 1997, when 55 Valencia orange harvesters became ill while working in groves that were treated with cyfluthrin 3–10 d earlier (Edmiston et al. 1998, 1999). Although the restricted entry interval was 12 hr, the registration status of cyfluthrin changed in April 1997 when the preharvest interval was reduced from 150 d to the day of harvest. The harvesters sought medical care for symptoms involving primarily respiratory irritation, including rhinitis (54.2%), sneezing (81.9%), coughing (21.7%), and sore throat (33.7%). Other symptoms included headache (21.7%), nausea (6.0%), and skin (16.9%) and eye irritation (16.9%). The symptoms were transitory, and most of the harvesters returned to work the following day.

B. Cyfluthrin Inhalation Monitoring Study

As part of the episode investigation, WH&S monitored the inhalation exposures of three experienced orange harvesters during 6 hr of harvesting (Edmiston et al. 1998). The monitoring began 30 hr after the Valencia orange grove was treated with cyfluthrin at the rate of 1 lb/A (0.45 kg/A). Cyfluthrin dust was trapped on 25-mm glass fiber filters housed in Institute of Occupational Medicine (IOM) samplers attached via tubing to personal air sampling pumps. Table 5 presents cyfluthrin residues found on the filters (mean = 5.13 μg/sample), inhalation concentration (mean = 7.13 μg/m³),

Table 5. Potential Inhalation Exposure for Orange Harvesters Exposed to Cyfluthrin Residues in California, 1997.

	Cyfluthrin residues (μg/sample)	Inhalation concentration (μg/m³)	Potential inhalation[a] (μg/kg)	Absorbed dosage[a] (μg/kg/day)
Worker 1	2.19	3.04	38.72	0.553
Worker 2	5.60	7.78	100.24	1.432
Worker 3	7.60	10.56	136.42	1.949
Mean	5.13	7.13	91.79	1.311

[a]Based on 8-hr exposure.

potential inhalation exposure (mean = 91.79 μg/kg, based on 8-hr exposure), and estimated absorbed dosage (mean = 1.311 μg/kg/d). Numerous cyfluthrin toxicology studies on file with DPR show irritant symptoms associated with a peak concentration of 100 μg/m³; this is also reported in the literature (Pauluhn and Machemer 1998). Adequate data to determine an acute no observable adverse effect level (NOEL) for human respiratory irritation following continuous exposures are not available. However, because the monitoring study found average inhalation concentrations provided a margin of exposure (MOE) of less than 100 fold compared to the experimentally established irritant threshold, DPR placed all products containing cyfluthrin into reevaluation in May 1998 (DPR, unpublished data). As part of the reevaluation process, DPR has required the primary registrant of cyfluthrin products to conduct and submit the results of inhalation irritation threshold studies and monitor corn harvesters exposed to cyfluthrin residues.

C. Cyfluthrin Dislodgeable Foliar Residue (DFR) Monitoring

WH&S sampled DFR from the episode groves and the grove where the inhalation monitoring study was conducted to determine average cyfluthrin residues during the monitoring period (Edmiston et al. 1999). To evaluate residue dissipation, WH&S sampled seven Valencia orange groves located within 50 mi of the episode groves over a 9-wk period following application. Pest control operators treated all sampled groves with cyfluthrin at 0.10 lb (0.05 kg) active ingredient/A in 100–250 gal (379–946 L) water. All tank mixes also included nutrients or buffers.

The following average cyfluthrin residues were found for the episode groves (6–12 d postapplication) and the grove where the inhalation monitoring study was conducted (11 d postapplication):

Episode groves, 0.039 ± 0.011 μg/cm²

Monitoring groves, 0.035 ± 0.007 μg/cm²

Using nonlinear regression analysis, the model $ln\,DFR = a \pm b * \sqrt{(days)}$ was the lowest order model that fit well for most of the fields. Table 6 displays the associated data for intercept, slope, and R^2 under this model. Initial deposition (as reflected by the model intercept) was similar across groves. The dissipation rates in the seven groves fell into two distinct decay patterns, with more rapid decay in groves 1–4 (overall average half-life = 4.9 d) and a considerably longer decay in groves 5–7. The half-life for groves exhibiting slower residue dissipation under the fitted model is not constant, as it would be with a first-order exponential decay model. Instead, each half-life is longer than the preceding one. The first two half-lives for groves 5–7 can be approximated as 11 and 32 d, respectively. In these groves, approximately 10%–20% of the initial residue was still present at 65 d

Table 6. Cyfluthrin Dislodgeable Foliar Residue (DFR) Dissipation in eight Valencia Orange Groves Treated with Cyfluthrin in California, 1997.[a]

Grove	Intercept	Slope	Average half-life (d)	R^2
1	-3.009	-0.397	5.73	0.63
2	-2.625	-0.441	4.82	0.81
3	-2.969	-0.450	4.67	0.75
4	-2.802	-0.470	4.37	0.89
5	-2.754	-0.233	19.6	0.67
6	-2.782	-0.258	14.7	0.69
7	-3.050	-0.152	95.6	0.55

[a]Fit of the model $ln\ DFR = a + b * \sqrt{(days)}$ *(days after application)*.
Source: Edmiston et al. (1999).

postapplication. Cyfluthrin appears to exhibit a variable dissipation rate, with the use of urea and/or potassium nitrate possibly associated with slower dissipation.

D. λ-Cyhalothrin Illness Episode

λ-Cyhalothrin, first registered in California in 1998, was the causal pyrethroid in 12 of the 317 illnesses previously discussed. Symptom arrays included irritant (5 cases), irritant/systemic (6 cases), and systemic (1 case). Six illnesses resulted from exposure to residues. Although the λ-cyhalothrin technical fact sheet states that adverse health effects include both skin and respiratory irritation, it documents only skin irritation as an adverse effect (National Pesticide Telecommunication Network 2004). The incidents cited involved laboratory personnel and workers who either handled concentrated λ-cyhalothrin and/or applied dilute λ-cyhalothrin solutions. The following describes an illness episode in which field workers developed acute respiratory symptoms on exposure to extremely high foliar residues of λ-cyhalothrin.

In September 1999, 11 raisin harvesters developed acute respiratory irritation symptoms when they entered a Fresno County vineyard (episode field) and were exposed to residues of λ-cyhalothrin, propargite, and sulfur (unpublished data, DPR). Within approximately 3 hr, crew members experienced sneezing, flu-like symptoms, and burning and itching on their arms, neck, face, and eyes. Seven workers received medical treatment. Subsequent investigation confirmed that the pesticide product Warrior Insecticide (U.S. EPA registration number 10182-00434-AA, containing 13.1% λ-cyhalothrin; Syngenta Crop Protection, Inc.), which was not registered for use on grapes, was mistakenly mixed and applied 45 d prior at 35 times the highest legal rate for any crop (Syngenta Crop Protection Inc. 2004). Gas chromatography analyses of eight DFR samples verified mean residues of λ-cyhalothrin ($0.43 \pm 0.10 \mu g/cm^2$), propargite ($0.35 \pm 0.11 \mu g/cm^2$), and sulfur ($0.31 \pm 0.28\,\mu g/cm^2$) on the grape leaves. The Fresno County Agricultural Commissioner declared the episode field a hazardous area due to high propargite and λ-cyhalothrin residues and prohibited worker reentry. WH&S compared episode DFR with 1998 λ-cyhalothrin residues for romaine lettuce and found episode field residues were approximately 12 times greater than those measured previously (Hernandez et al. 1998). The effects of exposure to average λ-cyhalothrin DFR levels of $0.43\,\mu g/cm^2$ have not been previously documented (Extension Toxicology Network 2004).

All eight DFR samples from the episode field showed propargite levels above $0.20\,\mu g/cm^2$, the estimated safe reentry level for repeated exposures to nectarine harvesters (O'Malley et al. 1990). Propargite and sulfur have been implicated in a number of dermal illness episodes over the past 20 yr (Gammon et al. 2001; Maddy et al. 1981; O'Malley et al. 1990; O'Malley 1998a; Winter and Kurtz 1985). Although propargite levels may have had a role in the incident, WH&S considered the high levels of λ-cyhalothrin to be the primary cause. Because information was unavailable for estimating the decay rate for such high residues, WH&S remediation for the episode field included posting the field against worker entry until natural defoliation had occurred and specifying that others who may enter before defoliation wear a properly fitted N95 particulate respirator, a disposable coverall, boots, and disposable gloves (DPR, unpublished data). The applicator was cited and paid a civil penalty of $1,000. λ-Cyhalothrin exposure resulting in illness was noted in a previous indoor applicator study, in which applications were made at legal rates (Moretto 1991).

V. Conclusions

In limited DPR monitoring studies associated with harvesting oranges at approximately two-thirds the normal work intensity, cyfluthrin air levels ($10\,\mu g/m^3$ cyfluthrin, measured as a time-weighted average) approached experimentally established irritant thresholds (LOELs) for airborne cyfluthrin ($100\,\mu g/m^3$ measured as an initial peak level) (Edmiston et al. 1998; Pauluhn and Machemer 1998). These observations and the common set of symptoms seen in experimental studies, PISP illnesses, and agricultural, structural, and indoor residue exposures reported in the literature suggest that field residues can cause irritant respiratory symptoms. It is unclear whether nonspecific symptoms such as nausea and headache reported in some of those indicate true systemic toxicity. Additional data are needed to establish threshold levels for both irritant and systemic symptoms for cyfluthrin and other pyrethroids.

For regulations based upon reentry intervals or waiting periods, additional dissipation data may also be necessary. Pyrethroid residues on crops, agricultural products, and foliage have previously been investigated, but pyrethroid dissipation is not well understood, as compared to the dissipation of organophosphorus pesticides (Argauer et al. 1997; Bellows et al. 1993; Dejonckheere et al. 1982; Edmiston et al. 1999; Estesen and Buck 1990; Giles et al. 1992; Hernandez et al. 1998; McEwen et al. 1986; Miyata et al. 1993; Nakamura et al. 1993; Papadopoulou-Mourkidou et al. 1989). Dissipation appears to vary significantly among pyrethroids and between crops, which hampers developing models to describe residue transfer in occupational settings. Previous studies of pyrethroid residues on cotton and orange foliage found half-lives of 5.3 and 6.2 d, respectively (Bellows et al. 1993; Estesen and Buck 1990). This result is similar to the dissipation exhibited in groves 1–4 in the WH&S cyfluthrin DFR study (see Table 6). Another study reported pyrethroid dissipation half-lives of 6.9–18.2 d for greenhouse-grown chrysanthemums, similar to the slower dissipation rate observed for groves 5 and 6 in the WH&S cyfluthrin DFR study (Giles et al. 1992). Dissipation as slow as that observed for grove 7 in WH&S' cyfluthrin DFR study (overall average half-life = 96 d; see Table 6) has not been reported in the literature. Dissipation following structural applications has been studied to only a limited extent, but sampling of carpet from treated homes in Germany has demonstrated that residues can persist for several years after application (Prohl et al. 1997). The Prohl study found that 83% of the subjects who had carpets removed had complete or partial improvement in nonspecific systemic and topical irritant symptoms compared to 16% of subjects who left treated carpet in place. These data are impressive even though the observational design of the study did not control for possible placebo effects of removing the treated carpet.

Summary

This survey summarizes California's recent experience with illnesses related to pyrethroid exposures and augments the data available on pyrethroid inhalation exposure and residue dissipation. We reviewed California Department of Pesticide Regulation (DPR) Pesticide Illness Surveillance Program (PISP) data and DPR Pesticide Use Reporting (PUR) data for 13 pyrethroids used during 1996–2002 and identified 317 illnesses associated with exposure. PUR found a total of 4,629,852 pounds (2,100,068 kg) of the 13 active ingredients were applied during the 7 yr. Type II pyrethroids accounted for 1,979,352 (897,820 kg) and 42.7% of the total pounds applied and 220 (69.6%) of the reported illnesses. Cyfluthrin was associated with 122 cases (55% of illnesses related to type II pyrethroids and 38.4% of all pyrethroid illnesses).

Agricultural uses accounted for 118 (37.3%) of the reported illness cases, with 116 cases associated with employment. For the 199 cases (62.8%) associated with nonagricultural use, 132 (66.3%) were occupationally related. Overall, approximately equal numbers of illnesses resulted from individual exposures (167 cases) and group exposures (150 cases). The symptom arrays associated with the pyrethroid illnesses included irritant effects or paresthesias of the eye, skin, or respiratory tract in 269 cases (84.9%). Type II pyrethroids were more frequently associated with isolated irritant symptoms (107 cases) than the type I pyrethroids (26 of 97 cases). Systemic symptoms were reported in 184 illnesses (58% of cases). Isolated systemic effects occurred in 48 cases (15.1%), but systemic effects were also present in 136 (50.6%) of the 269 cases with irritant symptoms. Residue exposures accounted for 158 illnesses (49.8%). Single or multiple violations of pesticide use regulations contributed to exposures in 90 of the 317 illnesses (28.4%); 76 were related to nonagricultural pyrethroid use.

We also report results of DPR Worker Health and Safety Branch (WH&S) investigations of three large group illness episodes related to exposure to type II pyrethroids cyfluthrin and λ-cyhalothrin that involved primarily respiratory irritation symptoms. An inhalation monitoring study found cyfluthrin air levels that approached experimentally established irritant thresholds for airborne cyfluthrin, from which a mean estimated absorbed dosage of 1.311 µg/kg/d was calculated. Although additional data are needed to establish threshold levels for both irritant and systemic symptoms for cyfluthrin and other pyrethroids, these observations suggest that field residues can cause irritant respiratory symptoms. DPR conducted a residue dissipation study in seven orange groves and estimated cyfluthrin residue half-lives. The dissipation rates fell into two distinct decay patterns, with more rapid decay in groves 1–4 (overall average half-life = 4.9 d) and a considerably longer decay in groves 5–7. The half-life for groves exhibiting the slower residue dissipation was not constant. The first two half-lives for groves 5–7 can be approximated; they are 11 and 32 d, respectively.

The third investigation involved an illness episode in which 11 raisin harvesters developed acute respiratory irritation symptoms when they were exposed to residues of λ-cyhalothrin, propargite, and sulfur. Gas chromatography analyses of eight dislodgeable foliar residue (DFR) samples verified mean residues of λ-cyhalothrin (0.43 ± 0.10 µg/cm^2), propargite (0.35 ± 0.11 µg/cm^2), and sulfur (0.31 ± 0.28 µg/cm^2) on the grape leaves. Subsequent investigation confirmed that the λ-cyhalothrin product, which was not registered for use on grapes, was mistakenly mixed and applied 45 d earlier at 35 times the highest legal rate for any crop. The effects of exposure to average λ-cyhalothrin DFR levels of 0.43 µg/cm^2 have not been previously documented.

References

Argauer, R, Eller, K, Pfeil, R, Brown, RT (1997) Determining ten synthetic pyrethroids in lettuce and ground meat by using ion-trap mass spectrometry and electron-capture gas chromatography. J Agric Food Chem 45:180–184.

Bellows, T, Morse, J, Gastin, L (1993) Residual toxicity of pesticides used for lepidopteran insect control on citrus to *Aphytix melinus debach* (Hymenoptera: Aphelinidae). Can Entomol 125:995–1001.

Cagen, SZ, Malley, LA, Parker, CM, Gardiner, TH, Van Gelder, GA, Jud, VA (1984) Pyrethroid-mediated skin sensory stimulation characterized by a new behavioral paradigm. Toxicol Appl Pharmacol 76:270–279.

DPR (California Department of Pesticide Regulation) Pesticide Use Report Data. http://www.cdpr.ca.gov/docs/pur/purmain.htm, accessed June 2004.

Casida, JE, Gammon, DW, Glickman, AH, Lawrence, LJ (1983) Mechanisms of selective action of pyrethroid insecticides. Annu Rev Pharmacol Toxicol 23:413–438.

Dejonckheere, W, Verstraeten, R, Steurbaut, W, Melkebeke, G, Kips, RH (1982) Permethrin and deltamethrin residues on lettuce. Pestic Sci 13:351–356.

Edmiston, S, Spencer, J, Hernandez, B, Fredrickson, AS (1998) Inhalation exposure of orange harvesters to cyfluthrin residue, HS-1765. California Department of Pesticide Regulation, Worker Health and Safety Branch. http://www.cdpr.ca.gov/docs/whs/pdf/hs1765.pdf, accessed June 2004.

Edmiston, S, Spencer, J, Hernandez, B, Fredrickson, AS (1999) Dissipation of dislodgeable foliar cyfluthrin residue, HS-1778. California Department of Pesticide Regulation, Worker Health and Safety Branch. http://www.cdpr.ca.gov/docs/whs/pdf/hs1778.pdf accessed June 2004.

Estesen, BJ, Buck, NA (1990) Comparison of dislodgeable and total residues of three pyrethroids applied to cotton in Arizona USA. Bull Environ Contam Toxicol 44:240–245.

Extension Toxicology Network (2004) EXTOXNET Pesticide Information Profiles, lambda-cyhalothrin. http://ace.orst.edu/info/extoxnet/pips/lambdacy.htm, accessed June 2004.

Fong, H (2001) Analysis of pesticide-related illnesses of select mixer/loader/applicators for years 1994 and 1998, HS-1824. California Department of Pesticide Regulation, Worker Health and Safety Branch. http://www.cdpr.ca.gov/docs/whs/pdf/hs1824.pdf accessed January 2005.

Gammon, D, Moore, T, O'Malley, M (2001) A toxicological assessment of sulfur as a pesticide. In: Krieger R (ed) Handbook of Pesticide Toxicology, 2nd Ed. Academic Press, San Diego, pp 1781–1791.

Giles, D, Blewett, T, Saiz, S, Welsh, AM, Krieger, RI (1992) Foliar and nontarget deposition from conventional and reduced-volume pesticide application in greenhouses. J Agric Food Chem 40:2510–2516.

He, F, Sun, J, Wu, Y, Yao, P, Wang, S, Liu, L (1988) Effects of pyrethroid insecticides on subjects engaged in packaging pyrethroids. Br J Ind Med 45:548–551.

He, F, Wang, S, Liu, L, Chen, S, Zhang, Z, Sun, J (1989) Clinical manifestations and diagnosis of acute pyrethroid poisoning. Arch Toxicol 63:54–58.

Hernandez, B, Spencer, J, Schneider, F, Welsh, A, Fredrickson, S (1988) A survey of dislodgeable pesticide residues on crop foliage at field reentry, 1994–1995. HS-1728. Sacramento, California, California Department of Pesticide Regulation, Worker Health and Safety Branch. http://www.cdpr.ca.gov/docs/whs/pdf/hs1728.pdf, accessed June 2004.

Kolmodin-Hedman, B, Swensson, A, Akerblom, M (1982) Occupational exposure to some synthetic pyrethroids (permethrin and fenvalerate). Arch Toxicol 50:27–33.

Kolmodin-Hedman, B, Akerblom, M, Flato, S, Alex, G (1995) Symptoms in forestry workers handling conifer plants treated with permethrin. Bull Environ Contam Toxicol 55:487–493.

Maddy, K, Smith, C, Estrella, L, Updike, D, Kilgore, S, Ochi, E, Au, C (1981) Occupational illnesses and injuries to field workers exposed to pesticide residue in California as reported to physicians in 1980, HS-940. California Department of Pesticide Regulation, Worker Health and Safety Branch. http://www.cdpr.ca.gov/docs/whs/pdf/hs940.pdf, accessed June 2004.

Maddy, K, Edmiston, S, Richmond, D (1990) Illness, injuries, and deaths from pesticide exposures in California 1949–1988. Rev Environ Contam Toxicol 114: 57–123.

McCarthy, S (2003) Assessing pesticide-related illnesses/injuries among California fieldworkers performing irrigation tasks, HS-1845. California Department of Pesticide Regulation, Worker Health and Safety Branch. http://www.cdpr.ca.gov/docs/whs/pdf/hs18454.pdf, accessed January 2005.

McEwen, F, Braun, H, Ritcey, G (1986) Residues of synthetic pyrethroid insecticides on horticultural crops. Pestic Sci 17:150–154.

Mehler, LN, O'Malley, MA, Krieger, RI (1992) Acute pesticide morbidity and mortality: California. Rev Environ Contam Toxicol 129:51–66.

Miyata, M, Kamakura, K, Hirahara, Y, Okomoto, K, Narita, M, Hasegawa, M, Koiguchi, S, Yamana, T, Tonogai, Y, Ito, Y (1993) Studies on simultaneous determination of 12 pyrethroid and 29 organophosphorus pesticides in agricultural products. J Food Hyg Soc Jpn 34:496–507.

Moretto, A (1991) Indoor spraying with the pyrethroid insecticide lambda-cyhalothrin: effects on spraymen and inhabitants of sprayed houses. Bull WHO 69:591–594.

Muller-Mohnssen, H (1999) Chronic sequelae and irreversible injuries following acute pyrethroid intoxication. Toxicol Lett 107:161–175.

Nakamura, Y, Tonogai, Y, Tsumura, Y, Ito, Y (1993) Determination of pyrethroid residues in vegetables, fruits, grains, beans, and green tea leaves: applications to pyrethroid residue monitoring studies. J AOAC Int 76:1348–1361.

National Pesticide Telecommunication Network (2004) Lambda-cyhalothrin fact sheet. http://npic.orst.edu/npicfact.htm, accessed June 2004.

O'Malley, M (1997) Clinical evaluation of pesticide exposure and poisonings. Lancet 349:1161–1166.

O'Malley, M (1998a) Irritant chemical dermatitis among grape workers in Fresno County, August 1995. HS-1741. California Department of Pesticide Regulation, Worker Health and Safety Branch. http://www.cdpr.ca.gov/docs/whs/pdf/hs1741.pdf, accessed June 2004.

O'Malley, M (1998b) Illnesses and injuries associated with exposures to phosphine and phosphine decomposition products. HS-1756. California Department of Pesticide Regulation, Worker Health and Safety Branch. http://www.cdpr.ca.gov/docs/whs/pdf/hs1756.pdf, accessed January 2005.

O'Malley, M, Verder-Carlos, M (2001) Illnesses related to exposure to metam-sodium byproducts in Earlimart, California in November, 1999. HS-1808. California Department of Pesticide Regulation, Worker Health and Safety Branch. http://www.cdpr.ca.gov/docs/whs/pdf/hs1808.pdf, accessed January 2005.

O'Malley, M, Smith, C, Krieger, R, Margetich, S (1990) Dermatitis among stone fruit harvesters in Tulare County, 1988. Am J Contact Dermatitis 1:100–111.

Papadopoulou-Mourkidou, E, Kotopoulou, A, Styliandis, D (1989) Field dissipation of the pyrethroid insecticide/acaricide biphenthrin on the foliage of peach trees, in the peel and pulp of peaches, and in tomatoes. Ann Appl Biol 115:405–416.

Pauluhn, J (1996) Risk assessment of pyrethroids following indoor use. Toxicol Lett 88:339–348.

Pauluhn, J (1999) Hazard identification and risk assessment of pyrethroids in the indoor environment. Toxicol Lett 107:193–199.

Pauluhn, J, Machemer, LH (1998) Assessment of pyrethroid-induced paraesthesias: comparison of animal model and human data. Toxicol Lett 96–97:361–368.

Pauluhn, J, Steffens, W, Haas, J, Machemer, L, Miksche, LK, Neuhauser, H, Shule, S (1996) Toxicologic evaluation of pyrethroids in indoor air: demonstrated with the example of cyfluthrin and permethrin. Gesundheitswesen 58:551–556.

Prohl, A, Boge, KP, Alsen-Hinrichs, C (1997) Activities of an environmental analysis van in the German Federal State Schleswig-Holstein. Environ Health Perspect 105:844–849.

Ray, D (2000) Pyrethroid insecticides: mechanisms of toxicity, systemic poisoning syndromes, paresthesia and therapy. In: Krieger R (ed) Handbook of Pesticide Toxicology, 2nd Ed. Academic Press, San Diego, pp 1289–1303, 2001.

Ray, DE, Forshaw, PJ (2000) Pyrethroid insecticides: poisoning syndromes, synergies, and therapy. J Toxicol Clin Toxicol 38:95–101.

Spencer J (2001) Analysis of the impact of the federal worker protection standard and recommendations for improving California's worker protection program regarding field posting, HS-1819. California Department of Pesticide Regulation, Worker Health and Safety Branch. http://www.cdpr.ca.gov/docs/whs/pdf/hs1819. pdf, accessed January 2005.

Syngenta Crop Protection, Inc. Specimen label number 099902 09-2590-065 R2 10/00, Warrior® Insecticide. http://www.cdms.net/manuf/, accessed June 2004.

U.S. EPA (1996) Food Quality Protection Act (FQPA) of 1996. Washington, DC, 2003. http://www.epa.gov/oppfead1/fqpa/backgrnd.htm#faca, accessed June 2004.

U.S. EPA (2000) Chlorpyrifos summary. Washington, DC, 2000. http://www.epa. gov/oppsrrd1/op/chlorpyrifos/summary.htm, accessed June 2004.

U.S. Government Accounting Office (1994) Pesticides on farms: limited capability exists to monitor occupational illnesses and injuries. Report to the chairman, Committee on Agriculture, Nutrition, and Forestry, US Senate GAO/PEMD-94-6.

Western Farm Press (2001) Farm Press University online. DPR releases data on 1999 pesticide injuries. http://www.westernfarmpress.com/news/farming_dpr_releases_data/, accessed January 2005.

Winter, CK, Kurtz, PH (1985) Factors influencing grape worker susceptibility to skin rashes. Bull Environ Contam Toxicol 35:418–426.

Manuscript received February 5; accepted February 7, 2005.

Rev Environ Contam Toxicol 186:73–105   © Springer 2006

Ecological Risk Assessment of Contaminated Soil

John Jensen and Marianne Bruus Pedersen

Contents

I. Introduction

Assessing the ecological risk of contaminated soil, pesticide application, sewage sludge amendment, and other human activities leading to exposure of the terrestrial environment to hazardous substances is a complicated task with numerous associated problems. Not only is terrestrial ecological risk assessment a relatively new field of science that has developed rapidly only since the mid-1980s, but it is also complicated by the fact that soil, in contrast to most aquatic environments, is very often on private lands and traded as real estate. Professional and economic divergence between the interests of scientists, stakeholders, authorities, engineers, managers, lawyers, non-government organizations (NGOs) and regulators is therefore not unusual. Even neglecting those aspects, a number of unresolved problems exist in

Communicated by George W. Ware.

J. Jensen (✉) · M.B. Pedersen
The National Environmental Research Institute, Department of Terrestrial Ecology, P.O. Box 314, Vejlsøvej 25, DK-8600 Silkeborg, Denmark.

the way we currently assess risk and manage the impact of anthropogenic substances in the terrestrial environment.

Ecological risk assessment (ERA) is a process of collecting, organizing, and analyzing environmental data to estimate the risk or probability of undesired effects on organisms, populations, or ecosystems caused by various stressors associated with human activities. The basic principles of ecological risk assessment are described in numerous papers (Ferguson et al. 1998; USEPA 1998; Suter et al. 2000; CSTEE 2000; Lanno 2003; Weeks et al. 2004; Thompson et al. 2005). All varieties of ERA are associated with uncertainties. The value or usefulness of the different ERA methodologies depends on the uncertainty, predictability, utility, and costs. There are typically two major types of ERA. The first is predictive and is often associated with the authorization and handling of hazardous substances such as pesticides or new and existing chemicals in the European Union. This kind of ERA is ideally done before environmental release. The second type of ERA could be described as an impact assessment rather than a risk assessment, as it is the assessment of changes in populations or ecosystems in sites or areas already polluted. The predictive method is based on more or less generic extrapolations from laboratory or controlled and manipulated semifield studies to real-world situations. The descriptive method is more site specific as it tries to monitor ecosystem changes in historically contaminated soils such as old dumpsites or gas facilities or in field plots after amendment with pesticides or sewage sludge, for example.

Often ERA is performed in phases or tiers, which may include predictive as well as descriptive methods. The successive tiers require, as a rule of thumb, more time, effort, and money. The paradigm or schemes for ERA may vary considerable from country to country, but often consist of an initial problem formulation based on a preliminary site characterization, and a screening assessment, a characterization of exposure, a characterization of effects, and a risk characterization followed by risk management. Although exposure assessment is often just as or even more important, this chapter primarily considers effect assessment.

In most European countries, ERA of contaminated soils consists of rather simplified approaches including soil screening levels (SSL) (a.k.a. quality objectives, quality criteria, benchmarks, guideline values) and simple bioassays for a first screening of risk. National research or remediation programs have led to the development of a large variety of guideline values. Although hard to categorize, most fall into two categories: generic or site specific. While the site-specific guidelines require a characterization of pH, organic matter, etc., at the site, generic guideline values are more independent of modifying factors and hence straightforward to legislate. Methodologies for deriving SSLs are described in Posthuma and Suter (2002) and Wagner and Løkke (1991).

Three major classes of tools for assessing ecological effects may be identified: standardized ecotoxicity experiments with single species exposed under controlled conditions to single chemicals spiked to soil; *ex situ* bio-assays, here defined as simple laboratory assays where single species are exposed to historically contaminated soils collected in the field; and finally monitoring, analyzing, and mapping of population or community structures in the field. Furthermore, mesocosm, lysometer, or terrestrial model ecosystems (TME) may be useful; these may be considered as large (multispecies) bioassays or ecotoxicity tests. TMEs have the advantage that they operate with the (relatively) undisturbed intrinsic soil populations that make up a small food web. TME hence allow the assessment of effects of toxicants that are mediated through changes in food supply or competition and predation. More information about TME is given in *Ecotoxicology* (vol. 13, no. 12, 2004).

One of the keystones in deriving environmental quality criteria is the use of standardized terrestrial test procedures. The emphasis of these prognostic tests is on reproducibility, standardization, international acceptance, and site independence. Although increasing in numbers, relatively few terrestrial tests are still approved by the International Standardisation Organisation (ISO) or Organization for Economic Cooperation & Development (OECD). However, other tests have shown promising results and are likely to be prepared for standardization in the future. A collection of terrestrial soil tests can be found in Tarradellas et al. (1997), Sheppard et al. (1992), Keddy et al. (1994), van Gestel and van Straalen (1994), DECHEMA (1995), and Løkke and van Gestel (1998). A list of guidelines can be found at www.iso.org and www.oecd.org.

However, the major problem in using simple laboratory tests to extrapolate to contaminated land may not be the limitations of test species and the natural variation in species sensitivity. The problems associated with extrapolating from one or a few species, exposed under controlled and typically optimal conditions, to the complex interaction of species and chemicals found in most contaminated ecosystems should also cause concern. Although single-species laboratory tests with spiked materials have their obvious benefits, e.g., they measure direct toxicity of chemicals and interpretation is therefore simple, supplementary tools are often needed.

Bioassays, as defined in this context (see above), are one of the more frequently used higher-tier alternatives. Basically the same test species may be used in bioassays for assessing the risk of a specific contaminated soil as in standard laboratory tests. However, bioassays have the advantage, compared to the use of spiked soil samples, that the exact toxicity of a specific soil may be assessed directly: this includes the combined and site-specific toxicological effect of the mixture of contaminants and their metabolites. Furthermore, the *in situ* bioavailability of that specific soil is (at least almost) maintained in the laboratory during the exposure period. Several studies have shown a reduction in bioavailability and/or toxicity of soils with an old history of contamination (Hatzinger and Alexander 1995; Alexander 1995, 2000; White and Alexander 1996; Kelsey and Alexander

1997; White et al. 1997; Tang et al. 1998; Robertson and Alexander 1998; Alexander and Alexander 2000; Morrison et al. 2000; Lanno et al. 2004). Bioassays are therefore often considered a more realistic tool than generic soil screening levels based on spiked laboratory soils. However, a number of uncertainties or problems may be associated with the use of bioassays and the interpretation of their results. First, the test species are still exposed to the contaminants in a relatively short period compared to the permanent exposure condition found at contaminated sites. Furthermore, they are exposed under more or less optimal conditions, in that stressors such as predation, inter- and intraspecies competition, drought, frost, and food depletion are eliminated during exposure. Finally, typically only a few species are tested individually.

To compensate for some of the limitations just described, contaminated soil may be assessed using multispecies mesocosms, lysometers, or TME. In these, species interactions may be evaluated by manually introducing several species to the systems or monitoring the intrinsic populations of the soil. Natural climatic conditions may be included if the test system is kept outdoors. However, if we want to get a more realistic and large-scale picture of the impact caused by, for example, pesticide use or sewage sludge application, or to assess the environmental health at waste sites, industrial areas, or gas works, it is often necessary to conduct some kind of field observations. Several case studies exist in which field studies have successfully elucidated the ecological risk of specific activities or the ecological impact at specific sites (Callahan et al. 1991; McLaughlin and Mineau 1995; Krüger and Scholtz 1998a,b; Holmstrup 2000; Salminen et al. 2001a,b; Kuznetsova and Patapov 1997; Ruzek and Marshall 2000).

The small single-species bioassay, large multispecies TME, and field surveys have some drawbacks in common. First of all, it may be difficult to actually link the observed effect to a specific toxic component in the soil. Which of the many substances is actually causing the majority of the observed effects, or is it perhaps a combination of effects? For a hazard classification of soils or a ranking of soils this may not be so important. However, to evaluate potential risk-reduction measures or risk-management procedures it may be important to identify the most problematic substances. A comparison of soil screening values with measured concentrations for each chemical present at a site may be helpful to identify the most likely group of substances causing the observed effect. Other possible tools may include a toxicity identification evaluation (TIE) approach (Burgess 2000; Carr et al. 2001; Babin et al. 2001; Van Sprang and Janssen 2001). The TIE approach is a relatively new method, which aims to identify groups of toxicants in soils with mixed pollution. Potentially toxic components present in the soil are fractionated and determined, and the toxicity of each individual fraction is determined by a *Lux* bacteria-based bioassay or the Microtox bioassay. Although perhaps promising, TIE is a time-consuming and hence costly procedure not yet used routinely.

Another crucial issue when analyzing the result of bioassays, TME, and field studies is the presence or absence of a proper reference site or soil. The control soil should in principle resemble the contaminated soil in all relevant parameters, e.g., texture, pH, organic matter, waterholding capacity, and nutrient content, a practical problem that very often is difficult to solve. The lack of adequate control or reference sites may, however, be conquered at least partially by the use of multivariate techniques (Kedwards et al. 1999a,b), which relate the species composition and abundance to gradients of pollutants. It is not the intention of this chapter to present a review of statistical tools for ecological risk assessment, and hence a detailed discussion about the use of these is not given. However, it is obvious that increased computer power and the presence of new easy-to-use software tools have increased the possibility to move away from more conventional univariate statistics such as analysis of variance (ANOVA) to more powerful multivariate statistics that use all collected data to evaluate effects at a higher level of organization. Statistical methods such as the power analysis may also be very useful in planning and designing large-scale ecotoxicity studies such as mesocosms, TME, or field surveys (Kennedy et al. 1999).

This chapter does not intend to present a comprehensive review of all published data from ecological studies at contaminated sites. Instead, three well-documented cases are presented to illustrate the described problems and challenges. Two cases, Cu and DDT, obtained from our own institute, The Department of Terrestrial Ecology at the Danish National Environmental Research Institute, are supplemented with studies from Dutch and UK zinc-contaminated soils. The Cu and Zn cases have been partly published in the open literature, whereas the DDT case is presented for the first time outside Denmark.

The observations from all three case studies are used in the discussion and form the basis for the final conclusion. In each case, we try to answer the following questions:

1. To what extent do soil screening levels (over)estimate risk?
2. Do bioassays represent a more realistic risk estimate?
3. Is it possible to make sound field surveys, or do we lack suitable reference situations?

II. Zinc-Contaminated Soils

Comparisons between field concentrations of the four most commonly studied metals (Cu, Pb, Cd, and Zn) and their relative toxicity to soil invertebrates have suggested that Zn, in a mixed metal pollution, is likely to be responsible for the majority of ecological effects observed (Spurgeon et al. 1994; Spurgeon and Hopkin 1995; Fountain and Hopkin 2004a,b). Therefore, in cases with mixed metal contamination, most attention has been paid to Zn.

A. Soil Screening Levels

The toxicity data on zinc in the literature are extensive and reveal a wide span in effect concentrations. Based on spiking experiments, the most sensitive species seem to be microorganisms or microbial processes, whereas soil invertebrates and plants are less sensitive (see following). The impact of zinc on microorganisms and microbial processes has been estimated in various ways e.g., impact on biomass, respiration, C/N mineralization, and enzyme activity. A few studies have reported no-effect values for microbial processes below or around $50 \, mg \, kg^{-1}$ with significant effects occurring from approximately $100 \, mg \, kg^{-1}$ (Wilson 1977; Tabatabai 1977; Bollag and Barabasz 1979). The lowest concentrations resulting in toxic effects on plants are about $100 \, mg \, kg^{-1}$ (MacLean 1974; Dang et al. 1990; Sheppard et al. 1993). For soil invertebrates, no-effect values range from approximately $200 \, mg \, kg^{-1}$ to several thousand milligrams per kilogram (Neuhauser et al. 1985; Spurgeon et al. 1994; Spurgeon and Hopkin 1995).

Using a relatively conservative approach, such countries as Denmark, The Netherlands, and Canada have established ecotoxicological soil quality standards for zinc. The Danish and the Canadian criteria in principal are based on the lowest fraction of the identified no-effect data, but the Dutch criteria are based on the national background level. Nevertheless, the criteria are comparable, i.e., 100, 200, and $140 \, mg \, kg^{-1}$ for the most sensitive land use in Denmark, Canada, and the Netherlands, respectively.

B. The Netherlands

A large Dutch research project with participation from Vrije University in Amsterdam, TNO, and National Institute for Public Health and Environmental Protection (RIVM) had as its major objective to evaluate the ecological relevance of toxicity data from laboratory studies and the risk limits derived from these data and to identify the factors introducing the largest uncertainties (Posthuma et al. 1998). Parts of the results have been internationally published by Posthuma et al. (1997), Smit et al. (1997, 2002), and Smit and van Gestel (1996, 1998). The project included laboratory tests with spiked soil and polluted field soil, spiked TME, and observations of communities of microorganisms, nematodes, and enchytraids in contaminated field soils. The field soil was located close to the smelter at Budel, The Netherlands. The results were numerous, and it is not possible to pay proper attention to all data in this review. The overall conditions and results from each category of test are presented below. More details can be found in the national report published in English (Posthuma et al. 1998).

Most tests both in the laboratory and in the field plots showed that bioavailability, and hence aging, was a dominant factor controlling toxicity of zinc. Regarding plant tests, the project concluded that results from experimental field plots using freshly and historically contaminated soils might be difficult to interpret. The effect data were strongly influenced by climatic factors; e.g., seed germination was only 9% in sunny areas whereas more than 80% of seed germinated in the shady areas. This factor, of course, hampered the correlation between laboratory studies and field studies as well as comparisons between years. It was therefore recommended to use controlled bioassays that are strictly limited to the vegetative phase instead of long-term field experiments when assessing phytotoxicity of contaminated soils. It is unclear whether a better test design using randomly block design and multivariate statistics could have improved the interpretation. Furthermore, they found that 0.01 M $CaCl_2$ extraction could be a useful tool for overcoming differences in bioavailability between field and laboratory and hence provided a better predictive tool for estimating phytotoxicity of zinc. The internal plant tissue concentration was properly also a useful indicator of effects, although the EC_{50} values expressed as the internal shoot tissue concentration varied by a factor of five between laboratory and field studies; however, so did the absolute growth rate of the plants. The higher growth in the field may have caused a dilution effect, resulting in lower internal concentrations in the shoot tissue. Although no information is available to support this conclusion, it may be postulated that concentrations at critical targets were similar.

The 0.01 M $CaCl_2$-extractable fraction of zinc was also a fairly good indicator of toxicity for springtails (*Folsomia candida*) when comparing spiked soil and historically contaminated soil from the field. Total concentrations, on the other hand, overestimated the toxicity of zinc to *F. candida* in all laboratory studies when compared to field-collected soil from a contaminated site and in artificially long term aged outdoor field plots. The EC_{50} for reproduction was between 184 and 626 mg Zn kg^{-1} for seven freshly spiked soils, but the EC_{50} for contaminated soil naturally aged for 18 mon was 2,178 mg Zn kg^{-1}. Both $CaCl_2$-extractable and water-soluble zinc were more appropriate estimates for toxicity than total concentrations. Equilibration of the zinc concentration by percolating the spiked soils with water before use in experiments strongly reduced the difference in toxicity between freshly spiked and aged soils.

The same test showed no negative effect of historically contaminated soil collected along a gradient from a smelter (Budel), although the highest concentration was 1,537 mg kg^{-1} (Smit and van Gestel 1996): this occurred although the water-soluble zinc concentrations found in the gradient samples from the three most contaminated cases exceeded the level, which reduced the reproductive output in the spiking experiments. Hence, not only was extractable zinc toxic, but other factors influenced toxicity of the investigated soils.

No fixed internal threshold concentration of zinc was observed for *F. candida*. Active metal regulation is common in many soil invertebrate species; therefore internal concentration, in contrary to plants, was not considered a valid tool for assessing the risk of zinc for soil microarthropods.

The Dutch research project also included tests with two species of oligochaetes, an earthworm (the compost worm *Eisenia andrei*) and an enchytraeid (the pot worm *Enchytraeus crypticus*). The tests included ordinary laboratory tests with spiked zinc, bioassays with contaminated field soil, and experimental field plots. Effects on the reproduction of earthworms were observed at significantly lower zinc concentrations when the worms were exposed to freshly spiked soils compared to historically contaminated field soil collected close to a smelter. Also, experimentally contaminated field soils spiked 2 and 3 yr before the test were less toxic to earthworms than freshly spiked test soils. They found that toxicity was pH dependent and, although bioavailability is not the only factor controlling toxicity, they concluded that the use of the 0.01 M $CaCl_2$-exchangeable fraction would reduce the uncertainty in extrapolating from laboratory to field conditions. Critical body concentrations in earthworms were relatively constant irrespective of the exposure conditions, and hence could be a possible alternative for total soil concentrations.

Analogous, but perhaps less profound, observations were made for the pot worm. Effect concentrations expressed as total concentrations varied by a factor of three or more between spiking experiments and bioassays with zinc-contaminated soil aged for 18 mon outside. When corrected for difference in bioavailability, comparable results were obtained in spiked laboratory studies and soils aged in outdoor plots. A historically contaminated soil collected from a gradient along a smelter showed toxicity comparable to freshly spiked soils, i.e., an EC_{50} of 205 mg kg^{-1} for the smelter soil and 262 mg kg^{-1} for the spiked soil. The fact that toxicity of the soil from the smelter area was higher that the two others indicates a mixed contamination or effects by other confounding stresses present in the smelter soil.

Field observations along a gradient from a smelter showed that the enchytraeid community was affected at the sites close to the factory, which was seen as reduced density and lower species numbers. However, other relevant anthropogenic changes, not unequivocally related to zinc or other metals, may also play a significant role in this. These factors include pH, private and military traffic, age of vegetation, and quantity and quality of organic matter, which all may have contributed to observed changes in soil conditions over short distances.

The effects of zinc on various nematode endpoints were also studied. Three months after spiked zinc contamination, the total nematode abundance showed a strong dose-related response in the field plots. After 10 mon, the abundance was normal in all but the three highest concentrations. This situation was maintained also after 22 mon. Comparison between more classical community endpoints such as total numbers of nematodes, taxonomic groups or species diversity, and principal response curves (PCR), which combines and incorporates all density data in a single analysis, showed that the latter was more sensitive.

Results for field-collected nematodes from a gradient around the smelter also turned out to be difficult to interpret. Zinc ranged from 1,787 mg Zn kg^{-1} near the smelter to 11 mg Zn kg^{-1} distant from the smelter. Stepwise log-logistic linear regressions were carried between biotic parameters and zinc content. The influence of soil factors was compensated by defining them as cofactors in the regression models. A statistical association between zinc and the number of species could not be demonstrated. Strong correlations between soil characteristics and the pollution gradient hampered a simple and straightforward comparison. The total number of nematodes was mainly related to moisture and CN ratio. The number of bacterial feeding nematodes was negatively correlated to pH and CN ratio. Only the plant parasitic nematodes showed a negative response to zinc. Furthermore, a strong dominance of a few species and the fact that the abundance of individual species fluctuated according to temperature, moisture, pH, and food availability made firm conclusions difficult.

On the basis of the field plot study and the smelter survey, it was concluded that nematode community characteristics such as diversity was a relative insensitive effect parameter, as many species, with a broad range of sensitivity, contribute to the community characteristics. Effect at overall community level is an average of effects on susceptible and more resistant species. Instead, response of individual species in the field was suggested as a toxicity endpoint. In this case, only 3 species, of a total of 35 identified species or 11 species with sufficient data, showed negative correlation to zinc concentrations.

Microbial studies revealed similar conclusions. First, the experimental field plot was found useful for studying effects of Zn on the mineralization of acetate and glutamate, for example. Second, the research group concluded that pollution-induced (microbial) community tolerance (PICT) could be a useful tool for assessing the environmental impact of soil contamination. Not only did they observe an increased tolerance at higher exposure levels in the experimental field plot, but they also found that the microbial community tolerance to zinc clearly decreased with increasing distance to the smelter.

C. United Kingdom

Since the mid-1990s, the group working with S.P. Hopkin at University of Reading, UK, have conducted a number of studies focusing on the effects of heavy metals, especially zinc, on soil invertebrates in the field and in the laboratory.

Spurgeon and Hopkin (1999a) did not find any increased heavy metal accumulation in worms exposed to spiked OECD soil compared to contaminated and aged field soil collected around a smelter in Avonmouth, located in the southwest of England. This result was not expected as years before they had demonstrated that toxic effects on earthworms (*Eisenia fetida*) of four heavy metals (Cu, Pb, Zn, and Cd) occurred at lower levels in spiked OECD soils than in seven field soils collected along the same gradient (Spurgeon and Hopkin 1995). Zinc was the most toxic metal, and the toxicity was found to be 10 times higher in artificial spiked soils than contaminated and aged field soils. The bioassays showed results comparable to the distribution of earthworms along the same gradient. Using bioassays in the laboratory, it was demonstrated that reproduction of earthworm was affected significantly when exposed to soil from the four sites closest to the smelter (Zn concentration, 2,793–32,871 mg kg^{-1}). No effects on reproduction or growth were observed when earthworms were exposed to soil collected from the three locations further away from the smelter; here, concentrations were 657–1,848 mg Zn kg^{-1}. However, the concentrations in these soils were still significantly higher than the lowest NOEC value (237 mg Zn kg^{-1}) established in laboratory studies using spiked OECD soil. The results from the bioassays in the laboratory correspond nicely with observations in the field. A significantly lower number of earthworms were found at the four sites closest to the heavy metal source compared to a nonpolluted site 110 km away. The zinc concentration at this site was 38 mg kg^{-1}. Compared to the nonpolluted site, a similar or higher number of earthworms was found at the three sites 3–7 km away from the smelter. However, the absence of significant effects on the total number of earthworms covers up the fact that species responded differently to heavy metal. *Apporectodea rosea* and *Allobophora chlorotica* both seemed sensitive to metal contamination, whereas *Lumbricus rubellus* and *Lumbricus castaneus* seemed less sensitive.

Spurgeon and Hopkin (1996) later expanded the field survey to include 22 sites within the affected area. Here they found *L. rubellus, L. castaneus*, and *L. terrestris* at sites close to the smelter whereas *Aporrectodea rosea*, *A. caliginosa*, and *Allolobophora chlorotica* were absent. Apparently *A. caliginosa* was the most affected species in the field as none was found in sites containing more than 900 mg Zn kg^{-1}. *L. castaneus*, on the other hand, was found in higher numbers close to the smelter. The results indicates that species-specific factors can be important in determining the responses of earthworm populations to metal pollution.

Toxicity tests in the laboratory with a "sensitive" and a "nonsensitive" earthworm species (*L. rubellus* and *A. rosea*, respectively) confirmed that *A. rosea* is more sensitive to zinc than *L. rubellus*. Thus, it appears that the distribution of species of earthworms around the smelter may be related to differences in sensitivity to zinc. Both species were, however, affected by zinc at lower concentrations than *Eisenia fetida*. The OECD test medium did not seem appropriate for *A. rosea* as this species did not reproduce normally in the control containers. The LOEC and EC$_{50}$ values for reproduction were 620 and 190 and 623 and 348 mg Zn kg^{-1} for *E. fetida* and *L. rubellus*, respectively. The compost worm is not found naturally in the field.

Hence, it cannot be confirmed whether this difference in sensitivity is valid also for field situations. Nevertheless, it demonstrates that the some margin of safety may be needed to extrapolate results from standard test organisms to field-relevant species.

In a later paper, Spurgeon and Hopkin (1999b) demonstrated that the dose–response relationship in *L. rubellus* collected from zinc-polluted areas such as the smelter in Avonmouth was only slightly different from worms collected in nonpolluted areas. Exposure to soil spiked with zinc in the laboratory resulted in only marginally higher reproductive output and similar mortality in the earthworms collected at the polluted sites. Hence, increased tolerance of earthworm strains exposed to zinc for hundreds of generations could not be demonstrated.

Fountain and Hopkin (2004b) investigated the effects of metal contamination on collembolans in the field and in the laboratory. They collected soil for laboratory bioassays and investigated species diversity and abundance of Collembola at 32 sampling points along a 35-m-long gradient of metal contamination in Ladymoor (Wolverhampton, England). The area was formerly used for disposal of smelting waste. Differences in the concentrations of Cd, Cu, Pb, and Zn between the least and most contaminated parts were more than one order of magnitude. The concentration at the sites was highly variable, i.e., from 597 to 9,080 mg Zn kg^{-1} (dry weight). In the laboratory they demonstrated that adult survival as well as number of juvenile *Folsomia candida* was significantly reduced when exposed to soils from Ladymor with approximately 8,000–9,000 mg Zn kg^{-1} (Fountain and Hopkin 2004a,b). When exposing the springtail to soils from all 32 sample sites, adult survival and reproduction showed significant negative relationships with total Zn concentrations. Survival was negatively correlated to total concentration ($R = 0.4033$, $P < 0.05$) and water soluble Zn concentrations ($R = 0.4271$, $P < 0.05$), whereas reproduction, unexpectedly, was correlated to total concentrations only ($R = 0.3438$, $P = 0.054$).

Springtails are generally not very sensitive to heavy metals due to their ability to excrete metals stored in the gut lining when they molt (Joosse and Buker 1979; Pawert et al. 1996). On the basis of the field survey, however, it was concluded that *Folsomia candida* was among the more sensitive springtails as it was found only at sites with concentrations below 2,300 mg Zn kg^{-1}. The other springtail regularly used in ecotoxicity testing, *F. fimetaria*, on the other hand is generally less sensitive to most heavy metals. *F. fimetarioides* is one of the dominant species found at heavy metal-polluted sites (Tranvik and Eijsacker 1989; Bengtsson and Rundgren 1988). At Ladymor it was also the dominant species at the highly contaminated plots, as it was mainly extracted from soil cores with concentrations between 5,000 and 10,000 mg Zn kg^{-1}. All in all, it was concluded that the *F. candida*, but not the *F. fimetaria*, bioassay was a useful tool to assess potential ecological impact at heavy metal-contaminated site as this species is considered relatively sensitive to zinc.

There was no obvious relationship between zinc levels and the total number of species or other commonly used diversity indices. However, individual species showed considerable differences in abundance. Tolerant species such as *Ceratophysella denticulata* and more sensitive species such as *Cryptopygus thermophilus* could be identified. Epedaphic (surface) species appeared to be less influenced by metal contamination than euedaphic (soil-dwelling) species. This difference probably results from the higher mobility and lower contact with the soil pore water of surface-dwelling springtails in comparison to soil-dwelling ones. The results showed the importance of considering effects also at the species level. Sensitive species that have important roles in the decompostion of organic material may be greatly reduced in numbers although there is no effect on the overall number of organisms.

In another study, Fountain and Hopkin (2004a) compared the effects of a highly contaminated soil from Ladymoor (total zinc, 7,907 mg kg^{-1}; water-extractable zinc, 18.4 mg kg^{-1}) with the effects of soil from four other contaminated areas. These all had lower metal concentrations (total and water-extractable zinc were in the range of 437–702 mg kg^{-1} and 3.0–10.3 mg kg^{-1}). The results of the bioassays with *F. candida* indicated highest risk of the Ladymoor soil. Nevertheless, the highest number of Collembola species was also found at the Ladymoor site, i.e., 24 compared to 15, 16, 16, and 19 at the other less contaminated sites. This result was partly explained by the fact that although Ladymoor has been a relatively undisturbed nature reserve almost unmanaged for 80 yr, the other sites were regularly disturbed by anthropogenic use. The paper did not present any data for relevant reference area to each of these contaminated sites. The fact that more species are found at Ladymoor compared to the four other sites therefore cannot exclude that adverse effects are occurring at Ladymoor. Neither can it be concluded that the lower number of species found at the less contaminated sites is a result of pollution. However, the observations demonstrate that the results from the *F. candida* bioassays cannot be used directly to predict species diversity at various sites or even to rank the sites according to total abundance of soil invertebrates.

Based on the foregoing, Fountain and Hopkin (2004a) suggested not using species diversity indices as a tool for interpreting metal contamination. They did not consider these indices as reliable estimates for risk of soil contamination, because the ranking of soils may change depending on the index used as the different indices put more or less emphasis on rare species, number of species, or number of individuals. They suggested instead the use of species composition as a more useful tool for assessing the risk of metal contamination and to focus on estimates of evenness and dominance instead of estimates of species richness. Ladymoor, for example, had the highest number of species, but was dominated by only two species (*Isotomurus palustris* and *Isotoma notabilis*), which is a typical characteristic for contaminated sites.

III. Copper-Contaminated Soil

For several years, the National Environmental Research Institute in Denmark conducted a number of biological investigations on a copper-contaminated site located in Hygum, Jutland, including a number of field studies. The site is a former wood preservation site that was operating in the period 1911–1924. It was cultivated from 1924 until 1993 and thereafter abandoned. In addition to the field studies we have performed a number of laboratory studies to underpin the observations made in the field. Most of the work has been published by Kjær and Elmegaard (1996), Bruus Pedersen et al. (1997), Scott-Fordsmand et al. (1997), Kjær et al. (1998), Bruus Pedersen et al. (1999, 2000a,b), Scott-Fordsmand et al. (2000a,b), and Bruus Pedersen and van Gestel (2001). Additionally, the Institute has derived the Danish soil quality criterion for copper and more than 20 other hazardous chemicals (Scott-Fordsmand and Jensen 2002).

A. Soil Screening Levels

Reviewing the toxicity data on copper in the literature reveals a wide span in effect concentrations. Based on simple spiking experiments, the most sensitive species to copper seem to be microorganisms, whereas soil invertebrates and plants are less sensitive. Effects on microbial endpoints have been observed from soil concentrations of 10–50 mg kg^{-1} (Maliszewska et al. 1985; Rogers and Li 1985), whereas sublethal effects on springtails, earthworms, and plants generally start above 100–200 mg kg^{-1} (Spurgeon et al. 1994; Kjær and Elmegaard 1996; Scott-Fordsmand et al. 2000a). As one exception to this, Scott-Fordsmand et al. (1997) reported effects on reproduction of springtails from approximately 40 mg kg^{-1} in spiked soil.

Using a relatively conservative approach, Denmark, The Netherlands, Canada, and other countries have established ecotoxicological soil screening levels (SSLs) or quality standards for copper and other metals. Although the Danish and the Canadian SSL in principle are based on the lowest fraction of the identified no-effect data, the Dutch criteria are based on the national background level of copper. Nevertheless, the criteria are comparable, i.e., 30, 63, and 36 mg kg^{-1}, respectively, for the most sensitive land use.

These criteria are established at concentrations considerably below the concentrations at which effects of Cu typically were observed in our laboratory. Bruus Pedersen and van Gestel (2001) found, for example, a 10% and 50% decrease in reproduction of springtails at 717 and 1,244 mg Cu kg^{-1}, respectively, when spiking reference soil from the field site at Hygum (Denmark). This level is a lower toxicity than observed by Scott-Fordsmand et al. (2000a), who found EC$_{10}$, EC$_{50}$, LC$_{10}$, and LC$_{50}$ values of 337, 994, 813, and 2,141 mg Cu kg^{-1}, respectively. In summary, based on the observation with spiked uncontaminated Hygum soil, reproduction of *F. fimetaria* may be affected from approximately $300\,mg\,Cu\,kg^{-1}$. This is a factor 10 fold higher than the quality criteria and almost 20 times higher than the background level of approximately $15\,mg\,Cu\,kg^{-1}$ found in that area.

It has long been recognized that the current algorithm for deriving soil quality standards has a limited use for heavy metals because naturally occurring background levels and the fact that many metals are essential for plants and soil organisms and hence actively regulated within the organisms. Furthermore, naturally occurring absorption processes will, with time, lead to a significant reduction in bioavailability, which may plead for an alternative approach for assessing environmental risk (Peijnenburg et al. 1997). Bruus Pedersen and van Gestel (2001) showed that increasing the aging time moderately between spiking and exposing springtails was, in the case of copper, not very useful for obtaining fieldlike conditions, as aging up to 12 weeks did not affect the results. It seems likely that the primary adsorption processes happen within the first hours after spiking and that further significant changes may take months or years. To elucidate the true effect of aging, field-collected soil samples (bioassays) or *in situ* monitoring of organisms therefore have to be used.

B. Bioassays

When exposing springtails (*F. fimetaria*) to concentrations of more than $2,500\,mg\,kg^{-1}$ in the laboratory, Bruus Pedersen and van Gestel (2001) and Scott-Fordsmand et al. (2000a) did not find any effects of soil collected along a gradient at the 70-yr-old copper-contaminated soil in Hygum. Exposing black bindweed (*Fallopia convolvulus*) to field-contaminated soil ($\leq 928\,mg\,Cu\,kg^{-1}$) had no effect on germination and only small effects on growth. Copper was considerably more phytotoxic when spiked into uncontaminated soil collected immediately next to the Hygum site. Here shoot growth was affected at dosages above $200\,mg\,Cu\,kg^{-1}$, reduced to 50% at $280\,mg\,Cu\,kg^{-1}$, and virtually absent above $400\,Cu\,kg^{-1}$ (Kjær et al. 1998; Bruus Pedersen et al. 2000b). The discrepancy between spiked soil and soil from the Cu-polluted site reflected that copper was less available to the plants as expressed by the reduced amount of extractable copper in the field soil. Extractions with distilled water and $0.01\,M$ $CaCl_2$, but also DTPA showed large disparity between spiked soil and field soil, especially at the higher end of the concentration range (Bruus Pedersen et al. 2000b; Scott-Fordsmand et al. 2000a; Bruus Pedersen and van Gestel 2001).

C. Field Studies

The effects of copper on field populations of microarthropods and plants at the Hygum site has been reported by Kjær et al. (1998), Bruus Pedersen et al. (1999), Scott-Fordsmand et al. (2000a,b), and Strandberg et al. (2005).

Scott-Fordsmand et al. (2000a) found that the populations of *F. fimetaria* responded differently to copper contamination depending on the season. In a May sampling no effects on the number of *F. fimetaria* could be observed even at the highest concentration. When animals were sampled in October, a relatively strong effect was observed, although not significant at the 5% level. Soil samples covering concentration ranges of 1,000–2,000 and 2,000–3,000 mg kg^{-1}, respectively, contained 70% and 80% less *F. fimetaria* compared to the number found in the range 15–1,000 mg kg^{-1}. By fitting a sigmoid distribution curve to the data, they calculated a field EC$_{10}$ level of approximately 640 mg kg^{-1}. The discrepancy between spring and fall data is explained by a far higher reproduction rate of springtails in the fall. Sampling at that time of the year is hence more likely to transmit possible effects on reproduction, and thus the population is more likely to be affected by lower copper concentrations. By taking more samples, monitoring more species, and using multivariate statistics, Bruus Pedersen et al. (1999) investigated the same site in more detail. The sampling was performed in the fall. Total microarthropod abundance was highest at intermediate concentrations. The Shannon–Wiener index of biodiversity decreased linearly with increasing copper concentrations, and the application of multivariate statistics showed that copper very well fitted the distribution of microarthropods in the field. A distinction in microarthropod community structure could be made between the soil samples with concentration between 50–200 mg kg^{-1} and those with higher concentrations. Although they also found that the species composition of the microarthropod community may be affected by other parameters, in this case the shade effects of a nearby row of trees, effect from copper was by far the most evident factor and it could be separated by other effects by using subsets of the total dataset. The authors concluded that multivariate analysis of community structure was a sensitive and useful method superior to single-species field data.

The performance of black bindweed (*F. convolvulus*) in the field was negatively correlated to both Cu concentrations and number of years the field has been a fallow field (Fig. 1) (Strandberg et al. 2005). In the first sampling year, 2 yr after cultivation was abandoned, *F. convolvulus* was found at Cu concentrations up to 496 mg Cu kg^{-1} in the part of the field site close to neighboring trees, whereas it only occurred below 60 mg Cu kg^{-1} in the open area away from the trees. This difference in occurrence over the field might have been an effect of the dense swards formed by the Cu-tolerant grass *Agrostis stolonifera* in the area far from the trees. In the following sampling years, *F. convolvulus* disappeared at lower and lower copper levels, and at the same time the perennial vegetation grew denser. This finding indicates an interactive effect of soil copper and interspecific competition. At any rate, the laboratory data obtained in experiments performed in copper-spiked soil appear more indicative of the field situation than laboratory tests in historically contaminated soil (bioassays).

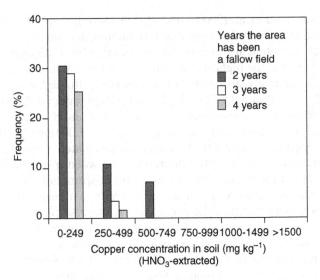

Fig. 1. Frequency of (*Fallopia convolvulus*) in three successive years at various ranges of copper concentrations at the Hygum site, Denmark.

IV. DDT-Contaminated Soils

On behalf of the Danish County Århus, The National Environmental Research Institute was requested to perform a site-specific risk assessment of a number of DDT-contaminated sites within the County. All sites were located in forest areas because the contamination was caused by massive use of DDT as insecticides in Christmas tree production. The results have so far only been published nationally in an internal report to the County. Therefore, some more details on sampling, methods, and dose–effect responses are included in here.

A. Soil Screening Levels

In Denmark, no SSLs for DDT exist. In the Netherlands, the target value is based on national background levels and is as low as $0.00009\,mg$ $kg^{-1}dw$. However, when testing for soil quality it is recommended to use a target value of $0.01\,mg\,kg^{-1}$ instead, which is extrapolated from ecotoxicity data. The Dutch intervention value, which indicates significant ecological risk, is $4\,mg\,kg^{-1}$. The Canadian quality guidelines for DDT in soil are set between 0.7 and $12\,mg\,kg^{-1}$, depending on the land use (agricultural/residential vs. commercial/industrial use). In both cases, the criterion is based on a relatively restricted number of toxicity data. Most of the existing data dealing with effects of DDT on soil dwelling species were derived in the 1960s and 1970s, with the consequence that data often are presented in a form not matching modern ecotoxicology, e.g., without $LC_{50}/EC_{50}/EC_{10}$ and

NOEC estimations, and therefore not directly useful in many national algorithms for setting environmental standards. Edwards and Thompson (1973) have reviewed the old data on DDT and soil fauna. A few of the more recent and suitable studies are reviewed below to give details to the existing quality standards. Denneman and van Gestel (1990, as cited in van der Plassche et al. 1994), observed LC_{50} values between $0.08\,mg\,kg^{-1}$ (quartz sand) and $77\,mg\,kg^{-1}$ (muck) when exposing first-instar nymphs of the species *Gryllus pennsylvanicus* to DDT for 18–24 hr in 12 different soils. The geometric mean of these 12 acute toxicity data ($10\,mg\,kg^{-1}$) forms the basis for the Dutch quality standard of $0.01\,mg\,kg^{-1}$, i.e., an assessment factor of 1,000 was used for the derivation of the SSL. Thomson and Gore (1972) tested the toxicity of 29 insecticides to the springtail *Folsomia candida* in a sandy soil, including DDT. They did not observe any significant mortality in the test concentration range ($0.005–50\,mg\,kg^{-1}$). The same was true for topical application, as no mortality was observed after spraying individuals with a 1% DDT solution. Robertson and Alexander (1998) did not observe any effect on the survival of three insect species at soil concentrations up to $600\,mg\,kg^{-1}$. However, at $2,500\,mg\,kg^{-1}$ all fruit flies (*Drosophila melanogaster*) and cockroaches (*Blattella germanica*) were killed, whereas it took $4,000\,mg\,kg^{-1}$ to kill all house flies (*Musca domestica*). Furthermore, Robertson and Alexander (1998) found that aging of DDT in the soil for 120 d significantly reduced the toxicity, as insect mortality was reduced from 100% to 68%, 52%, and 18% for cockroaches, fruit flies, and house flies, respectively. The decrease in toxicity was a combination of reduced bioavailability and degradation as approximately 85% of the applied DDT could still be extracted from the soil.

It seems reasonable, on the basis of the available toxicity test, to conclude that DDT has low toxicity toward soil-dwelling species and a SSL below $1\,mg\,kg^{-1}$ seems very precautionary and over protective.

B. Bioassays

DDT-contaminated soil was collected from seven field locations in Århus County, Denmark, and used in controlled bioassays in the laboratory. Acute tests with springtails and predatory mites showed low acute toxicity of DDT-contaminated soil. Five-day LC_{50} values for the springtail *Folsomia fimetaria* and the predatory mite *Hypoaspis aculeifer* were 6,500 (5,900–7,600) and 6,000 (−1,000–12,700) $mg\,DDT\,kg^{-1}$ dw, respectively (95% confidence interval in parentheses). The interpolated levels of no mortality (LC_{10}) were 1,100 (1,200–1,500) and 900 (200–9,100) $mg\,DDT\,kg^{-1}$ dw for the springtail and the mite, respectively. More information about the test methods can be found in Løkke and van Gestel (1998).

The reproduction of springtails was also relatively insensitive to DDT (Table 1). In many of the cases, only limited effects were observed even at the highest concentrations collected. At some sites large gaps between the

Table 1. Estimated No Observed Effect Levels (NOEC) and Lowest Observed Effect Levels (LOEC) ($P < 0.05$) for the Reproduction of the Springtail *Folsomia fimetaria* When Exposed to DDT-Contaminated Soils Collected Along Concentrations Gradients at Seven Danish Locations.

Location	NOEC	LOEC
Løvenholm	300	7,000
Mejlgård	157	460
Linå Vesterskov	195	>195
Sostrup	260	>260
Rødebækshus	74	>74
Hjøllund	>475	>475
Lyngsøhus	6,500	9,900

All data are expressed as mg DDT kg^{-1} (dw).

test concentrations were found. The presence of hot spots from dipping containers and spraying areas prevented a smooth concentration gradient at the sites. This is, nevertheless, a condition not restricted to these DDT-contaminated sites, but rather a common situation for many site-specific assessments. The DDT data demonstrate that, when exposed to historically contaminated field soil in the laboratory, springtail reproduction was generally not affected below 500 mg kg^{-1}. Based on data from all tests, an EC$_{10}$ value of 1,100 mg DDT kg^{-1} was calculated by the use of linear regression. In one of the soils (Lyngsøhus), the effects of DDT were studied not only on the number of offspring but also on the size of the juveniles. The latter seemed a more sensitive endpoint, as the NOEC and LOEC values were 1,500 and 3,000 mg DDT kg^{-1} versus 6,500 and 9,900 mg DDT kg^{-1} for the number of juveniles.

C. Field Studies

Most field studies have shown low toxicity of DDT to earthworms, enchytraids, nematodes, fleas, and springtails, whereas predatory mites were more sensitive (Wallace 1954; Edwards and Dennis 1960; Edwards et al. 1967). This finding was partly confirmed in our field survey. By analyzing 66 soil cores (0–5 cm deep with a Ø of 6 cm) from Løvenholm and 46 from Sostrup, we were able to estimate critical DDT levels for single collembolan species and for groups of mites, i.e., gamasites (predatory mites), prostigmates, and astigmates (Table 2). At both locations, we found approximately 15 springtail species, and mainly true soil-dwelling springtails such as *Tullbergia macrochaeta* and the prostigmates were reduced in numbers with increasing DDT levels. A few species, e.g., the springtail *Folsomia fimetaria*, increased in numbers, most likely as a result of the reduction in predatory mites. The number of *F. fimetaria* was more than doubled at DDT levels

Table 2. The Estimated EC_{10}, EC_{50}, NOEC, and LOEC ($P < 5\%$) Values for Three Groups of Mites [Gamasites (Predatory Mites), Prostigmates, and Astigmates] and the Most Common Microarthropod, Springtail *Tullbergia macrochaeta*, at Two DDT-Contaminated Sites in Denmark (Løvenholm and Sostrup).

Species	EC_{10}	EC_{50}	NOEC	LOEC
Løvenholm				
Gamasites	13.9 (−5.4–33.2)	91.5 (−35–218)	<100	100–300
Prostigmates	46.1 (−29.8–122)	303 (−196–802)	>200	>200
Astigmates	16.5 (−43–76)	108 (−278–495)	>200	>200
T. macrochaeta	6.5 (−5.0–18.1)	43 (33–119)	25–50	50–100
Sostrup				
Gamasites	0.12 (−0.11–0.35)	0.78 (−0.73–2.3)	<10	10–400
Prostigmates	0.44 (−0.13–1.02)	2.9 (−0.85–6.7)	25–175	175–400
T. macrochaeta	5.8 (−8.0–19.5)	38.1 (−52–129)	175–200	200–400

All values are in mg DDT kg^{-1} (dw).
Confidence intervals (95%) are indicated in parentheses.

above 100 mg kg^{-1}. At the Løvenholm site multivariate statistics, PRIMER (Clarke and Warwick 1994), were used to elucidate the effects of DDT on the microarthropod community. Populations of springtails and mites were evaluated against variables such as DDT levels, plant biomass, and organic matter content. Samples were pooled according to their DDT level (0–50, 50–100, 100–150, 150–200, and more than 200 mg DDT kg^{-1} dw). A significant difference in the composition of the microarthropod community was found between the group of samples with 0–50 mg DDT kg^{-1} and the other concentration ranges. Springtail species such as *Neelus minimus*, *Folsomia nana*, and *F. fimetaria* (stimulation) contributed most to the observed differences. Further subdivision of the 0–50 mg kg^{-1} group did not reveal any significant differences in this group.

DDT did not significantly affect the abundance and biomass of earthworms in the two areas (Løvenholm and Sostrup). The levels of DDT (+DDE and DDD) in earthworms collected from soil containing 10–100 ($n = 8$) and 1,000–7,000 mg DDT kg^{-1} ($n = 3$), were 41 ± 26 and 86 ± 26 mg kg^{-1}, respectively. The levels of lindane (γ-HCH) in the earthworms were 0.012 ± 0.010 and 0.035 ± 0.014 mg kg^{-1}. Based on simple bioaccumulation models, these levels did not indicate any significant threat to worm-eating animals. This result was confirmed by analyses of liver tissue from two mice (*Sorex araneaus*) and two moles (*Talpa europaea*) containing 4.7/38.2 and 3.31/1.06 mg DDT kg^{-1} wet weight, respectively.

V. Discussion and Conclusions

The previous sections have described three cases of ecological risk assessment. The cases include two heavy metals (Cu and Zn) and an anthropogenic organic chemical (DDT). As an overall conclusion, it seems there are two major constraints hampering the use of laboratory tests to predict effects under natural field conditions. One key issue is bioavailability; another one is suboptimal conditions or multiple stresses in the field such as climatic stress (drought, frost, etc.), predators, competition, and food shortage. Although bioavailability for long has been recognized as a major problem in ERA (Chung and Alexander 2002; Morrison et al. 2000), the problem of how to incorporate multiple stresses, other than chemical mixtures, so far has been considered by only a few researchers (Holmstrup et al. 2000). The first problem, bioavailability, may partly be solved by conducting bioassays, at least in cases where the soil does not have to be modulated (pH adjusted or moistened) to make it fit for the test species. The problem about multiple stresses, on the other hand, is not elucidated by the use of bioassays but is inherently included in field surveys.

Bioavailability is mainly controlled by chemical and physical parameters in the soil such as pH, Cation-exchange Capacity (CEC), and organic carbon as well as by aging processes (Sandifer and Hopkin 1996; Smit and van Gestel 1996; van Gestel and van Diepen 1997; Bruus Pedersen et al. 1997; Chung and Alexander 2002). In the case of both Zn and Cu it was obvious that the difference in bioavailability, and hence also toxicity, between naturally aged contamination and freshly spiked contamination was large. Both cases, however, also demonstrated that the bioavailability of heavy metals could be mimicked reasonable well by extraction with weak solvents such as $CaCl_2$ or (rain)water (Bruus Pedersen and van Gestel 2001; Smit et al. 1997). Although not yet well established, similar work is going on within the field of estimating the bioavailable fraction of organic compounds. Extraction procedures such as butanol, tetrahydrofuran, ethanol, C18 membranes, solid-phase microextraction, and sequential fluid extraction have shown potential for predicting the bioavailability of organics and hence perhaps also toxicity (Kelsey et al. 1997; Ramos et al. 1998; Tang and Alexander 1999; Berglof et al. 2000; Tang et al. 2002; Hartonen et al. 2002). Nevertheless, much work still has to be done before a conceptual framework for including bioavailability can be implemented in environmental risk assessment and risk management (Peijnenburg et al. 1997; Lanno 2003). First, it must be recognized that bioavailability is governed by dynamic processes comprising several distinct phases: one is the physicochemically driven adsorption/desorption process (chemical availability) controlled by parameters such as pH, clay, CEC, and organic matter, another is a physiologically driven uptake process (biological availability) controlled by species-specific parameters such as anatomy, feeding strategy, and preferences in microhabitat, and the last is an internal allocation process (toxicological availability) controlled by species-specific parameters such as metabolism, detoxification, storage, excretion, or energy resources. Internal concentrations are hence most relevant when target organs have been identified and their critical levels estimated. Although chemical availability is very site specific, biological and toxicological availability are generally less site specific, although physiological and genetic adaptation processes at contaminated sites may influence the species-specific uptake and allocation processes.

Until these processes and their mutual relationship are better elucidated, it must be recommended to be careful when incorporating bioavailability in ERA. It should, nevertheless, be considered to include essential parameters controlling chemical availability such as pH and clay for heavy metals and organic matter for organic compounds in the derivation of soil quality standards. Varying the quality standards according to these parameters would increase the environmental relevance. Some countries, such as the Netherlands, have partly done this. On the other hand, it should be emphasized that we know only little about biological and toxicological availability. Furthermore, bioavailability is not the only factor that significantly varies between laboratory and field. As shown in the previous sections, other factors may result in higher toxicity in the field than in controlled laboratory studies. The use of quality standards as the first tier should therefore still be based on a large degree of conservatism to not undermine the precautionary principle.

Another recurring observation in most of the presented cases was that much attention must be paid to choosing proper reference situations, i.e., control soils in bioassays and reference sites in field studies. The reference soil has to resemble the contaminated soil in essential parameters such as pH, moisture, nutrient content, and organic matter quantity and quality, and preferably reference sites have to be similar to the contaminated site when it comes to slope, illumination, and land management. These factors seem especially important for plant studies. To a certain extent, it may be necessary to manipulate the control soil of bioassays to reach the same pH, nutrient status, or organic matter content as the contaminated soil. However, this should be done with caution, and it is not advised to change the pH of the contaminated soil, for example, to optimize it for the test species, as it is likely to change bioavailability and hence most likely also toxicity. In cases where the physicochemical conditions in the test soil are out of range for the optimal condition of the test species, an alternative test species must be considered. For plant tests, it is recommended to add excess nutrients to minimize the difference between confounding effects in the two sets of soils. This step will furthermore most likely amplify a potentially toxic response in the plants and hence increase the sensitivity of the plant test (Redente et al. 2002; Stephenson et al. 2001).

The examples presented of copper, DDT, and also partly zinc show that caution should be paid to the interpretation of the results with bioassays. In some cases, bioassays did not report effects on the selected test species at levels at which marked effects were found in the field. In the case of copper no effects on the test species, the springtail *Folsomia fimetaria* and the plant *Fallopia convolvulus*, were observed even when they were exposed to soil from the hot spots. Nevertheless, in the field the same species was clearly affected in the hot spots. The microarthropod and the plant community as a whole were affected at even lower concentrations than the individual species. In the case of DDT, bioassays with the springtail *F. fimetaria* showed a very low toxicity in most test soils. Nevertheless, monitoring in the field revealed a clear effect on springtail species, and the microarthropod community as a whole, at relatively low concentrations. It should, however, be emphasized that the field study indicated that the species used in the laboratory studies, *F. fimetaria*, seemed relatively insensitive to DDT compared to other springtail species and mites. Data from the copper and zinc sites also indicated that *F. fimetaria* was not the most sensitive springtail species to heavy metals. These observations plead for caution when interpreting the results of bioassays using only one species.

One explanation for the observation of apparently higher toxicity in the field compared to testing the same soil in the laboratory may be confounding effects of chemical stress and competition, reduced food availability, predation, and/or climatic stress in the field. Holmstrup and coworkers have shown synergism between climatic and chemical stress (Holmstrup 1997; Holmstrup et al. 1998, 2000; Højer et al. 2001; Sjursen et al. 2001). There is no easy solution to this problem unless an additional safety measure somehow is applied to the output of bioassays. But how do you add an additional safety factor in cases where no effects are observed even at the highest concentrations? Another solution would be to include a combination of stresses in the laboratory tests. Although an interesting research topic, it is not recommendable for the purposes of risk management, to perform standard laboratory studies under suboptimal conditions as these must change according to the test species and the test soil in question. Although more realistic, results from laboratory tests will be less transparent if performed under combined stresses. Instead, it may be recommended to focus on field studies.

There are a number of important issues to address when planning and conducting a field study. One aspect is the choice of a proper reference site (Attrill and Depledge 1997). Furthermore, it is important to take advantage of modern statistical computer tools when planning the experiments and analyzing the collected data. Power analyses may indicate how many samples are needed to detect changes under the specific conditions, and the application of multivariate statistics makes it fairly straightforward to analyze changes in community structure as an alternative to changes in single-species populations. The former has been argued by many to be superior to the latter (Joern and Hoagland 1996; Korthals et al. 1996; Maund et al. 1999). The presented examples with zinc show how large natural fluctuations hamper a firm conclusion of field studies when looking at the total abundance instead of the overall structure of soil communities or the development of single indicator species. A solid foundation in single-species toxicity studies is, however, always beneficial as these may help interpreting results from the field. The outcome of these simplified, but strongly controlled, experiments may furthermore help to identify the potentially most hazardous substances in a toxic mixture as the ones observed at many contaminated sites.

There is generally scientific consensus about the need of a tiered approach in ecological risk assessment of contaminated soil. It is also commonly agreed that some kind of generic soil screening levels are needed as a first tier. Assessing higher tiers of ecological risk assessment should, however, contain some kind of site-specific assessment. The examples presented here clearly demonstrate that a simple use of bioassays, i.e., one or two ecotoxicological tests with site-specific soil samples, may not be sufficient to capture ecological effects actually happening at the specific site for a number of reasons. One factor is the duration of exposure, which is relatively short compared to the lifelong exposure that organisms often experience at a contaminated site. Another may be the exclusion of confounding stresses such as drought, cold, food depletion, competition, and predators. Furthermore, bioavailability may change when soil samples are handled in the laboratory. It is therefore strongly recommended to look carefully at the results of bioassays and to include field studies in cases of reasonable doubt. Terrestrial model systems (TME) or similar multispecies tests may help elucidating the confounding stress associated with food web interactions. TMEs will furthermore enable shortcutting some of the discussion about how to protect soil structure or soil function (Van Straalen 2003), as it is possible to measure structure as well as essential soil functions in the model ecosystems.

In addition to new and better test systems, it is obviously important for stakeholders that the ERA procedure is constructed in a logical way that enables the risk assessor to give answers to some relevant questions: What kind of effects are ecologically, scientifically, or socially acceptable? Is it acceptable that the ecosystem structure has changed so long as the overall functioning is maintained, e.g., one springtail species has increased in numbers at the cost of three others? What are the (long-term) ecological consequences of a 28% reduction in the reproductive output of a bioassay with one species? Who is to decide what is acceptable and when?

It is therefore of highest importance, before the start of the risk assessment, to have consensus among all stakeholders, authorities, and risk assessors about these questions: What are we trying to protect in this specific case (target of protection)? Is it ecosystem structure, specific ecological functions or species, or a combination of both? How may these be linked? How will we assess the risk, i.e., which "surrogates" may we choose to represent our targets of protection? What are the acceptable criteria in the selected tests, e.g., NOEC or EC_{50}? Do we aim at the most sensitive 5% of the species sensitivity distribution (SSD)? Questions such as these need to be addressed before a strong, coherent, and transparent ecological risk assessment procedure can be conducted. Finally, it may often be necessary to consider how to establish the pathway between source of pollution and the chosen receptors.

It is of highest importance to organize the various studies in a framework or decision support system that is transparent and useful for all stakeholders. Uncertainty will exist no matter what kind of tests and risk assessment procedure are chosen. A weight of evidence approach may be an obvious choice to deal with these uncertainties. The sediment quality TRIAD, as developed by Chapman (Chapman 1986; Chapman et al. 1997, 2002) and proposed for the soil environment by Rutgers et al. (2000) and Rutgers and den Besten (2005), is a tool for dealing with conceptual uncertainties. The TRIAD systems acknowledge that all kind of test systems, i.e., chemistry, laboratory studies, bioassays, and field monitoring, may give us valuable information. All the information can be categorized in a triangle – chemistry, toxicology, and ecology – and if the scores exceed prescribed thresholds action can be taken. These thresholds may vary according to the intended land use. Furthermore, the different "legs" of the triad can be organized in different tiers increasing in complexity. The results of the various tests must be integrated in the risk assessment. Therefore, they have to be quantified in a way that allows all results to enter a decision matrix (Rutgers and den Besten 2005), which can be done by scaling all chemical, toxicological, and ecological results from 0 to 1 where 0 is no effect and 1 is the maximum or very severe effect. How to scale each parameter or test result is a matter of expert judgement, and this may be a challenge for some tests. Hereafter, one integrated effect value for each triad leg is calculated by the geometric mean. The integrated results for each leg are computed into one final risk assessment figure, including the risk deviation showing the imbalance between the results in the different legs of the triad. Large risk numbers and/or risk deviation argue for further studies. Cases where field data indicate changes in soil communities compared to an apparent reference site but chemical analyses and bioassays show no risk at all may, for example, trigger further studies. The same may be true the other way around. Work is continuously going on in developing and refining existing frameworks for ecological risk assessment (Weeks et al. 2004). The TRIAD approach seems at this moment to be one of the most appropriate ways to reduce the uncertainty in ecological risk assessment in a pragmatic way.

Summary

This review has described three cases of ecological risk assessment. The cases include two heavy metals (Cu and Zn) and an anthropogenic organic chemical (DDT). It concludes that there are at least two major constraints hampering the use of laboratory tests to predict effects under natural field conditions. One key issue is bioavailability, and another is suboptimal conditions or multiple stresses in the field such as climatic stress (drought, frost), predators, competition, or food shortage.

On the basis of the presented case studies, it was possible to answer three essential questions often raised in connection to ecological risk assessment of contaminated sites.

1. *To what extend does soil screening level (SSL) estimate the risk?*
The SSL are generally derived at levels corresponding to the lowest observed effect levels in laboratory studies, which often is close to the background levels found in many soils. In the cases of zinc and especially DDT, the SSL seemed quite conservative, whereas for copper they resemble the level at which changes in the community structure of soil microarthropods and the plant community have been observed at contaminated sites. The SSL correspond as a whole relatively well with concentrations where no effects or only minor effects were observed in controlled field studies. However, large variation in field surveys can often make it difficult to conclude to what extent the SSL corresponded to no-effect levels in the field.

2. *Do bioassays represent a more realistic risk estimate?*
Here, there is no firm conclusion. The zinc study in UK showed a better relationship between the outcome of *ex situ* bioassays and field observations than the SSL. The latter overestimated the risk compared to field observations. However, this would be species dependent, as the sensitivity to metals may vary considerably between recognized test species, even within the same group of organisms, such as *Folsomia candida* and *Folsomia fimetaria* or *Eisenia fetida* and *Lumbricus sp.* Furthermore, it was demonstrated that bioassays were not useful for predicting general species diversity in the field as they are strongly influenced by natural variation and other factors not related to contamination. In the case of copper, bioassays with springtails and black bindweed seemed to underestimate the risk compared to the Cu concentrations at which significant changes in the community structure of soil microarthropods and plants have been observed at the contaminated site, and this was also the case for the DDT-contaminated soils. Here, bioassays with DDT-contaminated soils showed generally very low toxicity, with EC_{10} values considerably higher than the levels where clear effects on single species as well as community structure have been detected in the present field study.

3. *Is it possible to make sound field surveys or do we lack suitable reference situations?*
Large natural variation caused by other factors than contaminants were observed in most cases, and this may have particularly hampered the conclusions made in the field surveys. These factors included pH, private and military traffic, age of vegetation, shading effects, and variations in light insensitivity as well as quantity and quality of organic matter. It was therefore concluded that field studies should always be interpreted in concert

with similar data from a reference situation. Conclusions should therefore be made with caution in situations where important soil conditions vary between control plots and the contaminated sites. The cases also showed that indices focusing on species richness were unreliable. Estimates of evenness or dominance were recommended instead, and most authors concluded that multivariate analysis of community structure was a sensitive and useful method superior to single-species field data.

This review concludes that there is a need for a tiered approach in ecological risk assessment of contaminated soils. Generic soil screening levels are needed as a first tier. Higher tiers of ecological risk assessment should, however, contain some kind of site-specific assessment. It is furthermore important to organize the various studies in a framework or decision support system that is transparent and useful for all stakeholders. A weight of evidence approach may be an obvious choice to deal with these uncertainties. The TRIAD approach, which incorporates and categorizes information in a triangle – chemistry, toxicology, and ecology – is an appropriate tool for handling conceptual uncertainties.

Acknowledgments

Financial support from the European Research Project "Development of a Decision Support System for Sustainable Management of Contaminated Land by Linking Bioavailability, Ecological Risk and Ground Water Pollution of Organic Pollutants" (LIBERATION) (Contract no. EVK1-CT-2001-00105) is acknowledged. The work by colleagues Paul Henning Krogh, Martin Holmstrup, and Jørgen Axelsen on the DDT case is greatly appreciated.

References

Alexander, M (1995) How toxic are toxic chemicals in soil? Environ Sci Technol 29:2713–2717.

Alexander, M (2000) Aging, bioavailability, and overestimation of risk from environmental pollutants. Environ Sci Technol 34:4259–4265.

Alexander, RR, Alexander, M (2000) Bioavailability of genotoxic compounds in soils. Environ Sci Technol 34:1589–1593.

Attrill, MJ, Depledge, MH (1997) Community and population indicators of ecosystem health: Targeting links between levels of biological organisation. Aquat Toxicol 38:183–197.

Babin, MM, Garcia, P, Fernandez, C, Alonso, C, Carbonell, G, Tarazona, JV (2001) Toxicological characterisation of sludge from sewage treatment plants using toxicity identification evaluation protocols based on in vitro toxicity tests. Toxicol In Vitro 15:519–524.

Bengtsson, G, Rundgren, S (1988) The Gusum case: a brass mill and the distribution of soil Collembola. Can J Zool 66:1518–1526.

Berglof, T, Koskinen, WC, Kylin, H, Moorman, TB (2000) Characterization of triadimefon sorption in soils using supercritical fluid (SFE) and accelerated solvent (ASE) extraction techniques. Pest Manag Sci 10:927–931.

Bollag, JM, Barabasz, W (1979) Effects of heavy metals on the denitrification process in soil. J Environ Qual 8:196–201.

Bruus Pedersen, M, van Gestel, CAM (2001) Toxicity of copper to the collembolan *Folsomia fimetaria* in relation to the age of soil contamination. Ecotoxicol Environ Saf 49(1):54–59.

Bruus Pedersen, M, Temminghoff, EJM, Marisussen, MPJC, Elmegaard, N, van Gestel, CAM (1997) Copper accumulation and fitness of *Folsomia candida* Willem in a copper contaminated sandy soil as affected by pH and soil moisture. Appl Soil Ecol 6:135–146.

Bruus Pedersen, M, Axelsen, JA, Strandberg, B, Jensen, J, Attrill, MJ (1999) The impact of a copper gradient on a microarthropod field community. Ecotoxicology 8:467–483.

Bruus Pedersen, M, van Gestel, CAM, Elmegaard, N (2000a) Effects of copper on reproduction of two Collembolan species exposed through soil, food, and water. Environ Toxicol Chem 19:2579–2588.

Bruus Pedersen, M, Kjaer, C, Elmegaard, N (2000b) Toxicity and bioaccumulation of copper to black bindweed (*Fallopia convolvulus*) in relation to bioavailability and the age of soil contamination. Arch Environ Contam Toxicol 39:431–439.

Burgess, RM (2000) Characterizing and identifying toxicants in marine waters: a review of marine toxicity identification evaluations (TIEs). Int J Environ Pollut 13:2–33.

Callahan, CA, Menzie, CA, Burmaster, DE, Wilborn, DC, Ernst, T (1991) On-site methods for assssesing chemical impact on the soil environment using earthworms: a case study at the Baird and McGuire superfund site, Holbrook, Massachusetts. Environ Toxicol Chem 10:817–826.

Carr, RS, Nipper, M, Biedenbach, JM, Hooten, RL, Miller, K, Saepoff, S (2001) Sediment toxicity identification evaluation (TIE) studies at marine sites suspected of ordnance contamination. Arch Environ Contam Toxicol 41:298–307.

Chapman, PM (1986) Sediment quality criteria from the sediment quality triad: An example. Environ Toxicol Chem 3:957–964.

Chapman, PM, Anderson, B, Carr, S, Engle, V, Green, R, Hameedi, J, Harmon, M, Haverland, P, Hyland, J, Ingersoll, C, Long, E, Rodgers, J, Salazar, M, Sibley, PK, Smith, PJ, Swartz, RC, Thompson, B, Windom, H (1997) General guidelines for using sediment quality triad. Mar Pollut 34:368–372.

Chapman, PM, McDonald, BG, Lawrence, GS (2002) Weight-of-evidence issues and frameworks for sediment quality (and other) assessments. Hum Ecol Risk Assess 8:1489–1515.

Chung, N, Alexander, M (2002) Effect of soil properties on bioavailability and extractability of phenanthrene and atrazine sequestered in soil. Chemosphere 48:109–115.

Clarke, KR, Warwick, RM (1994) Similarity-based testing for community pattern – the 2-way layout with no replication. Mar Biol 118:167–176.

CSTEE (Scientific Committee on Toxicity, Ecotoxicity and the Environment) (2000) The available scientific approaches to assess the potential effects and risk of chemicals on terrestrial ecosystems. Report from the European Commission, C2/JCD/csteeop/Ter91100/D(0), Brussels, Nov. 9, 2000.

Dang, YP, Chhabra, R, Verma, KS (1990) Effect of Cd, Ni, Pb and Zn on growth and chemical composition of onion and fenugreek. Commun Soil Sci Plant Anal 21:717–735.

DECHEMA (1995) Bioassays for soil. Methods for toxicological/ecotoxicological assessment of soils (Kreysa G, Wiesner J, eds) Frankfurt A.M., Germany.

Denneman, CAJ, van Gestel, CAM (1990) Bodemveronreiniging en bodemecosystemen: Voorstel voor C- (toetsings)waarden op basis van ecotoxicologisch risico's. RIVM report no. 725201 001. National Institute for Public Health and Environmental Protection (RIVM).

Edwards, CA, Dennis, EB (1960) Some effects of aldrin and DDT on the soil fauna of arable land. Nature (Lond) 188:767.

Edwards, CA, Thompson, AR (1973) Pesticides and the soil fauna. Residue Rev 45:1–79.

Edwards, CA, Dennis, EB, Empson, DW (1967) Pesticides and soil fauna: Effects of aldrin and DDT in an arable field. Ann Appl Biol 60:11.

Ferguson, C, Darmendrail, D, Freier, K, Jensen, BK, Jensen, J, Kasamas, H, Urzelai, A, Vegter, J, (eds) (1998) Risk Assessment for Contaminated Sites in Europe, vol 1. Scientific Basis. LQM Press, Nottingham, United Kingdom.

Fountain, MT, Hopkin, SP (2004a) Biodiversity of Collembola in urban soils and the use of Folsomia candida to assess soil "quality." Ecotoxicology 13:555–572.

Fountain, MT, Hopkin, SP (2004b) A comparative study of the effect of metal contamination on Collembola in the field and in the laboratory. Ecotoxicology 13:573–587.

Hartonen, K, Bowadt, S, Dybdahl, HP, Nylund, K, Sporring, S, Lund, H, Oreld, F (2002) Nordic laboratory intercomparison of supercritical fluid extraction for the determination of total petroleum hydrocarbon, polychlorinated biphenyls and polycyclic aromatic hydrocarbons in soil. J Chromatogr A 958:239–248.

Hatzinger, PB, Alexander, M (1995) Effect of aging of chemicals in soil on their biodegradability and extractability. Environ Sci Technol 29:537–545.

Holmstrup, M (1997) Drought tolerance in Folsomia candida Willem (Collembola) after exposure to sublethal concentrations of three soil-polluting chemicals. Pedobiologia 41:361–368.

Holmstrup, M (2000) Field assessment of toxic effects on reproduction in the earthworms Aporrectodea longa and Aporrectodea rosea. Environ Toxicol Chem 19:1781–1787.

Holmstrup, M, Petersen, BF, Larsen, MM (1998) Combined effects of copper, desiccation, and frost on the viability of earthworm cocoons. Environ Toxicol Chem 17:897–901.

Holmstrup, M, Bayley, M, Sjursen, H, Højer, R, Bossen, S, Friss, K (2000) Interactions between environmental pollution and cold tolerance of soil invertebrates: A neglected field of research. Cryo-Letters 21:309–314.

Højer, R, Bayley, M, Damgaard, CF, Holmstrup, M (2001) Stress synergy between drought and a common environmental contaminant: studies with the collembolan Folsomia candida. Global Change Biol 7:485–494.

Joern, A, Hoagland, KD (1996) In defense of whole-community bioassays for risk assessment. Environ Toxicol Chem 15:407–409.

Joosse, ENG, Buker, JB (1979) Uptake and excretion of lead by litter-dwelling Collembola. Environ Pollut 18:235–240.

Keddy, C, Greene, JC, Bonnell, MA (1994) A review of whole organism bioassays for assessing the quality of soil, freshwater sediment and freshwater in Canada. Scientific series no. 198. Environment Canada, Ottawa, Ontario, Canada.

Kedwards, TJ, Maund, SJ, Chapman, PF (1999a) Community level analysis of eco-toxicological field studies: I. Biological monitoring. Environ Toxicol Chem 18:149–157.

Kedwards, TJ, Maund, SJ, Chapman, PF (1999b) Community level analysis of eco-toxicological field studies: II. Replicated-design studies. Environ Toxicol Chem 18:158–166.

Kelsey, J, Alexander, M (1997) Declining bioavailability and inappropriate estimation of risk of persistent compounds. Environ Toxicol Chem 16:582–585.

Kelsey, JW, Kottler, BD, Alexander, M (1997) Selective chemical extractants to predict bioavailability of soil-aged organic chemicals. Environ Sci Technol 31:214–217.

Kennedy, JH, Ammann, LP, Waller, WT, Warren, JE, Hosmer, AJ, Cairns, SH, Johnson, PC, Graney, RL (1999) Using statistical power to optimize sensitivity of analysis of variance designs for microcosms and mesocosms. Environ Toxicol Chem 18:113–117.

Kjær, C, Elmegaard, N (1996) Effects of copper sulfate on black bindweed (*Polygonum convolvulus* L.). Ecotoxicol Environ Saf 33:110–117.

Kjaer, C, Bruus, Pedersen M, Elmegaard, N (1998) Effects of soil copper on black bindweed (*Fallopia convolvulus*) in the laboratory and in the field. Arch Environ Contam Toxicol 35(1):14–19.

Korthals, GW, Alexiev, AD, Lexmond, TM, Kammenga, JE, Bongers, T (1996) Long-term effects of copper and pH on the nematode community in an agro-ecosystem. Environ Toxicol Chem 15:979–985.

Krüger, K, Scholtz, CH (1998a) Changes in the structure of dung insect communities after ivermectin usage in a grassland ecosystem. I. Impact of ivermectin under drought conditions. Acta Oecol 19:425–438.

Krüger, K, Scholtz, CH (1998b) Changes in the structure of dung insect communities after ivermectin usage in a grassland ecosystem. II. Impact of ivermectin under high-rainfall conditions. Acta Oecol 19:439–451.

Kuznetsova, NA, Patapov, MB (1997) Changes in structure of communities of soil springtails (Hexapoda: Collembola) under industrial pollution of the south Taiga Bilberry pine forrest. Russ J Ecol 28:386–392.

Lanno, RP (2003) (ed) Contaminated Soils: From Soil-Chemical Interactions to Ecosystem Management. SETAC Press, Pensacola, FL.

Lanno, RP, Wells, J, Conder, J, Bradham, K, Basta, N (2004) The bioavailability of chemicals in soil for earthworms. Ecotoxicol Environ Saf 57:39–47.

Løkke, H, van Gestel, CAM, (eds) (1998) Handbook of Soil Invertebrate Toxicity Tests. Wiley, & New York.

MacLean, AJ (1974) Effects of soil properties and amendments on the availability of zinc in soils. Can J Soil Sci 54:369–378.

Maliszewska, W, Dec, S, Wierzbicka, H, Wozniakowska, A (1985) The influence of various heavy-metal compounds on the development and activity of soil micro-organisms. Environ Pollut A 37:195–215.

Maund, S, Chapman, P, Kedwards, T, Tattersfield, L, Matthiessen, P, Warwick, R, Smith, E (1999) Application of multivariate statistics to ecotoxicological field studies. Environ Toxicol Chem 18:111–112.

McLaughlin, A, Mineau, P (1995) The impact of agricultural practices on biodiversity. Agric Ecosyst Environ 55:201.

Morrison, DE, Robertson, BK, Alexander, M (2000) Bioavailability to earthworms of aged DDT, DDE, DDD and dieldrin in soil. Environ Sci Technol 34:709–713.

Neuhauser, EF, Loehr, RC, Milligan, DL, Malecki, MR (1985) Toxicity of metals to the earthworm *Eisenia fetida*. Biol Fertil Soils 1:149–152.

Pawert, M, Triebskorn, R, Graff, S, Berkus, M, Schulz, J, Köhler, HR (1996) Cellular alterations in collembolan midgut cells as a marker of heavy metal exposure: ultrastructure and intrecellular metal distribution. Sci Total Environ 181:187–200.

Peijnenburg, WJGM, Posthuma, L, Eijsacker, HJP, Allen, HE (1997) A conceptual framework for implementation of bioavailability of metals for environmental management purposes. Ecotoxicol Environ Saf 37:163–172.

Posthuma, L, Suter, GW, (eds) (2002) The Use of Species Sensitivity Distributions (SSD) in Ecotoxicology. SETAC Press, Pensacola, FL.

Posthuma, L, Baerselman, R, Van Veen, RPM, Dirven-Van Bremen, EM (1997) Single and joint effects of copper and zinc on reproduction of *Enchytraeus crypticus* in relation to sorption of metals in soils. Ecotoxicol Environ Saf 38:108–121.

Posthuma, L, van Gestel, CAM, Smit, CE, Bakker, DJ, Vonk, JW (1998) Validation of toxicity data and risk limits for soils: final report. RIVM report no. 607505004. National Institute of Public Health and the Environment, pp 230.

Ramos, EU, Meijer, SN, Vaes, WHJ, Verhaar, HJM, Hermens, JLM (1998) Using solid-phase microextraction to determine partition coefficients to humic acids and bioavailable concentrations of hydrophobic chemicals. Environ Sci Technol 32:3430–3435.

Redente, EF, Zadeh, H, Paschke, MW (2002) Phytotoxicity of smelter-impacted soils in southwest Montana, USA. Environ Toxicol Chem 21:269–274.

Robertson, BK, Alexander, M (1998) Sequestration of DDT and dieldrin in soil: Disappearance of acute toxicity but not the compounds. Environ Toxicol Chem 17:1034–1038.

Rogers, JE, Li, SW (1985) Effect of metals and other inorganic ions on soil microbial activity: soil dehydrogenase assay as a simple toxicity test. Bull Environ Contam Toxicol 34:858–865.

Ruzek, J, Marshall, VG (2000) Impacts of airborne pollutants on soil fauna. Annu Rev Ecol Syst 31:395–423.

Rutgers, M, den Besten, P (2005) Approaches to legislation in a global context. The Netherlands perspective – soil and sediment. In: Toxicity Testing Environmental. Thompson C, Wahhia K, Loibner AP (eds). Blackwell, CRC Press, Oxford, pp 269–289.

Rutgers, M, Faber, JH, Postma, JF, Eijsackers, H (2000) Site-Specific Ecological Risks: A Basic Approach to the Function-Specific Assessment of Soil Pollution. The Netherlands Integrated Research Programme, Vol. 28.

Salminen, J, van Gestel, CAM, Strommer, R (2001a) Pollution-induced community tolerance and functional redundancy in a decomposer food web in metal-stressed soil. Environ Toxicol Chem 20:2287–2295.

Salminen, J, Anh, BT, van Gestel, CAM (2001b) Enchytraeids and microbes in Zn polluted soil: No link between organism-level stress responses and ecosystem functioning. Ecotoxicology 10:351–361.

Sandifer, RD, Hopkin, SP (1996) Effects of pH on the toxicity of cadmium, copper, lead and zinc to *Folsomia candida* Willem, 1902 (Collembola) in a standard laboratory test system. Chemosphere 33:2475.

Scott-Fordsmand, JJ, Jensen, J (2002) Ecotoxicological soil quality critaria in Denmark. In: Posthuma L, Suter GW (eds) The Use of Species Sensitivity Distributions (SSD) in Ecotoxicology. SETAC Press, Pensacola FL.

Scott-Fordsmand, JJ, Krogh, PH, Weeks, JM (1997) Sublethal toxicity of copper to a soil-dwelling springtail (*Folsomia fimetaria*) (Collembola: Isotomidae). Environ Toxicol Chem 16:2538–2542.

Scott-Fordsmand, JJ, Krogh, PH, Weeks, JM (2000a) Responses of *Folsomia fimetaria* (Collembola: Isotomidae) to copper under different soil copper contamination histories in relation to risk assessment. Environ Toxicol Chem 19:1297–1303.

Scott-Fordsmand, JJ, Weeks, JM, Hopkin, SP (2000b) Importance of contamination history for understanding toxicity of copper to earthworm *Eisenia fetida* (Oligochaeta: Annelida) using neutral-red retention assay. Environ Toxicol Chem 19:1774–1780.

Sheppard, SC, Gaudet, C, Sheppard, MI, Cureton, PM, Wong, MP (1992) The development of assessment and remediation guidelines for contaminated soils: A review of the science. Can J Soil Sci 72:359–394.

Sheppard, SC, Evenden, WG, Abboud, SA, Stephenson, M (1993) A plant life-cycle for contaminated soil for comparison with other bioassays: Mercury and zinc. Arch Environ Contam Toxicol 25:27–35.

Sjursen, H, Sverdrup, LE, Krogh, PH (2001) Effects of polycyclic aromatic compounds on the drought tolerance of *Folsomia fimetaria* (Collembola, Isotomidae). Environ Toxicol Chem 20:2899–2902.

Smit, C, van Gestel, CAM (1996) Comparison of the toxicity of zinc for the springtail *Folsomia candida* in artificially contaminated and polluted field soils. Appl Soil Ecol 3:127–136.

Smit, C, van Gestel, CAM (1998) Effects of soil type, percolation, and ageing on bioacccumulation and toxicity of zinc for the springtail *Folsomia candida*. Environ Toxicol Chem 17:1132–1141.

Smit, CE, van Beelen, P, van Gestel, CAM (1997) Development of zinc bioavailability and toxicity for the springtail *Folsomia candida* in an experimentally contaminated field plot. Environ Pollut 98:73–80.

Smit, CE, Schouten, AJ, Van den Brink, PJ, van Esbroek, MLP, Posthuma, L (2002) Effects of zinc contamination on a natural nematode community in outdoor soil mesocosms. Arch Environ Contam Toxicol 42:205–216.

Spurgeon, DJ, Hopkin, SP (1995) Extrapolation of the laboratory-based OECD earthworm test to metal-contaminated field sites. Ecotoxicology 4:190–205.

Spurgeon, DJ, Hopkin, SP (1996) The effects of metal contamination on earthworm populations around a smelting works: quantifying species effects. Appl Soil Ecol 4:147–160.

Spurgeon, DJ, Hopkin, SP (1999a) Comparisons of metal accumulation and excretion kinetics in earthworms (*Eisenia fetida*) exposed to contaminated field and laboratory soils. Appl Soil Ecol 11:227–243.

Spurgeon, DJ, Hopkin, SP (1999b) Tolerance to zinc in populations of the earthworm *Lumbricus rubellus* from uncomtaminated and metal-contaminated ecosystems. Arch Environ Contam Toxicol 37:332–337.

Spurgeon, DJ, Hopkin, SP, Jones, DT (1994) Effects of cadmium, copper, lead and zinc on growth, reproduction and survival of the earthworm *Esenia fetida* (Savigny): Assessing the environmental impact of point-source metal contamination in terrestrial ecosystems. Environ Pollut 84:123–130.

Stephenson, GL, Feisthauser, NC, Koper, N, McCann, JH, Scroggins, RP (2001) The influence of four types of water on seedling emergence and growth of barley and the toxic interaction with copper sulfate. In: Environmental Toxicology and Risk Assessment: Science, Policy and Standardisation: Implications for Environmental Decisions, vol 10. ASTM STP 1403. American Society for Testing and Materials, Washington, DC.

Strandberg, B, Axelsen, JA, Bruus, Pedersen M, Jensen, J, Attrill, MJ (2006) The impact of a copper gradient on plant community structure. Environ Toxicol Chem 25(3).

Suter, GW, Efroymson, RA, Sample, BE, Jones, DS (2000) Ecological Risk Assessment for Contaminated Sites. CRC Press Lewis, Boca Raton, FL.

Tang, JX, Alexander, M (1999) Mild extractability and bioavailability of polycyclic aromatic hydrocarbons in soil. Environ Toxicol Chem 18:2711–2714.

Tang, JX, Carroquino, MJ, Robertson, BK (1998) Combined effect of sequestration and bioremediation in reducing the bioavailability of polycyclic aromatic hydrocarbons in soil. Environ Sci Technol 32:3586–3590.

Tang, JX, Liste, HH, Alexander, M (2002) Chemical assays of availability to earthworms of polycyclic aromatic hydrocarbons in soil. Chemosphere 48:35–42.

Tarradellas, J, Bitton, G, Rossel, D (1997) Soil Ecotoxicology. CRC Lewis, Boca Raton, FL.

Tabatabai, MA (1977) Effects of trace metals on urease activity in soils. Soil Biol Biochem 9:9–13.

Thompson, AG, Gore, FL (1972) Toxicity of twenty-nine insecticides to *Folsomia candida*: Laboratory studies. J Econom Entomol 65:1255–1260.

Thompson, KC, Wahhia, K, Loibner, AP, (eds) (2005) Environmental Toxicity Testing. Blackwell, CRC Press, Oxford.

Tranvik, L, Eijsacker, H (1989) On the advantage of *Folsomia fimetarioides* over *Isotomiella minor* (Collembola) in a metal polluted soil. Oecologia (Berl) 80:195–200.

USEPA (1998) Guidelines for ecological risk assessment. Report from the U.S. Environmental Protection Agency/Risk Assessment Forum, EPA/630/R-95/002F. Fed Regis 63(93):26846–26924 (May 14, 1998).

van der Plassche, EJ, Canton, JH, Eijs, YA, Everts, JW, Janssen, PJCM, van Koten-Vermeulen, JEM, Polder, MD, Posthumus, R, de Stoppelaar, JM (1994) Towards integrated environmental quality objectives for several compounds with a potential for secondary poisoning: Underlying data. Annex to RIVM report no. 679101 012. National Institute for Public Health and Environmental Protection (RIVM), pp 130.

van Gestel, CAM, van Diepen, AMF (1997) The influence of soil moisture content on the bioavailability and toxicity of cadmium for *Folsomia candida*. Ecotoxicol Environ Saf 36:123–132.

van Gestel, CAM, van Straalen, NM (1994) Ecotoxicological test systems for terrestrial invertebrates. In: Donker M, Eijsackers H, Heimbach F (eds) Ecotoxicology of Soil Organisms, Lewis CRC Press, Boca Raton, FL, pp 205–228.

Van Straalen, NM (2003) Ecological stress becomes stress ecology. Environ Sci Technol 37(17):324A–330A.

Van Sprang, PA, Janssen, CR, (2001) Toxicity, identification, of, metals:, Development, of toxicity identification fingerprints. Environ Toxicol Chem 20:2604–2610.

Wallace, MMH (1954) The effect of DSDT and BHC on the population of the lucerne flea *Sminthurus viridis* (Collembola), and its control by predatory mites *Biscircus* ssp. (Bdellidae). Aust J Agric Res 5:148–1954.

Wagner, C, Løkke, H (1991) Estimation of ecotoxicological protection levels from NOEC toxicity data. Water Res 25:1237–1242.

Weeks, J, Sorokin, N, Johnson, IJ, Whitehouse, P, Ashton, D, Spurgeon, D, Hankard, P, Svendsen, C (2004) Biological test methods for assessing contaminated land. Stage 2: A demonstration of the use of a framework for the ecological risk assessment of land contamination. Science Group Report P5-069/TR1. Environment Agency, United Kingdom.

White, JC, Alexander, M (1996) Reduced biodegradability of desorption-resistant fractions of polycyclic aromatic hydrocarbons in soil and aquifer solids. Environ Toxicol Chem 15:1973–1978.

White, J, Kelsey, JW, Hatzinger, PB, Alexander, M (1997) Factors affecting sequestration and bioavailability of phenanthrene in soils. Environ Toxicol Chem 16:2040–2045.

Wilson, DO (1977) Nitrification in three soils amended with zinc sulphate. Soil Biol Biochem 9:277–280.

Manuscript received May 25, 2005; accepted May 27, 2005.

Rev Environ Contam Toxicol 186:107–132   © Springer 2006

Pesticide Exposure Monitoring Databases in Applied Risk Analysis

J.H. Ross, J.H. Driver, C. Lunchick, C. Wible, and F. Selman

Contents

Communicated by George W. Ware.

J.H. Ross (✉)
infoscientific.com, Inc., 5233 Marimoore Way Carmichael, CA 95608, U.S.A.

J.H. Driver
infoscientific.com, Inc., 10009 Wisakon Trail Manassas, VA 20111, U.S.A.

C. Lunchick
Bayer CropScience, 2 Alexander Drive Research Triangle Park, NC 27709, U.S.A.

C. Wible
The Scotts Company, 14111 Scottslawn Road Marysville, OH 43041, U.S.A.

F. Selman
Dow AgroSciences, 306/B2810 9330 Zionsville Road Indianapolis, IN 46268, U.S.A.

I. Introduction

The use of pesticides to control insects, fungi, microorganisms, weeds, and other pests is an integral part of modern agricultural and public health practices. Because pesticides are biologically active agents, there is concern for potential health risks incurred by people directly involved in their use and by persons who may experience collateral, i.e., reentry worker and bystander exposure. To quantify the potential health risks associated with pesticide use, it is important to understand not only the toxicicological profile and associated effects of concern for a particular active ingredient but also the pathways, routes (dermal, inhalation, incidental ingestion), magnitude, frequency, and variance of anticipated exposures. Populations routinely exposed to pesticides include agricultural handlers involved in cultivation of field crops, greenhouse crops, vineyards, and orchards; professional grounds applicators (e.g., parks, roadsides); lawn care professionals; structural and commercial applicators (e.g., for factories, food processing plants, hotels, hospitals, other institutions, offices, residences); field workers, e.g., during harvesting or canopy management of treated crops; and household residents (EPA 1984; Maddy et al. 1990).

Starting in the 1950s, research to quantitatively assess pesticide exposure was driven by evidence of occasional overexposure observed among workers. However, concern has been increasingly driven by prospective risk assessments with the goal of preventing excessive exposures. This paradigm shift in risk assessment/management was formalized in 1983 with the publication of the "Red Book" (NAS 1983). Although associative or causal relationships between human illness and observed laboratory animal toxicity (which frequently serves as the basis for the dose–response metric included in a risk assessment) are often not observed, public health protective measures endeavor to prevent illness or an unacceptable probability of an adverse effect. Further, regulatory safety standards continue to evolve, typically in more precautionary directions, in part due to advancements in measurement technologies, i.e., the ability to measure ever smaller exposures and characterize ever more sensitive biological changes. Therefore, an increased demand for more robust quantitative exposure analyses and the associated underlying data has resulted. This increased demand has prompted an ongoing need for cost-effective and scientifically sound generic exposure monitoring databases. Generic exposure data are derived from a monitoring study using a specific chemical, but the exposure data are broadly applicable to a variety of chemicals because exposure is a physical process and chemicals with a broad range of physicochemical properties will produce similar exposure under similar physical exposure conditions (Whitmyre et al. 2001; Krieger et al. 1992).

II. Task Force Responsibilities

There are a limited number of subject matter experts in the discipline of worker and consumer pesticide exposure monitoring and assessment. An examination of the published literature regarding worker exposure to pesticides, for example, reveals a list of names numbering only in the hundreds. International meetings of these experts have been hosted by the Dutch (van Hemmen et al. 1995, 2001), Canadian (Worgan and Rosario 1995), and U.S. government agencies and the International Life Sciences Institute (ILSI 2003). The economics of expensive data development programs are also an obvious reason for task force efforts. Economies of scale can be realized through collaborative task force efforts. Thus, task forces representing multiple companies, and composed of a collection of industry and consulting experts, are formed to procure and produce broadly applicable exposure-related data. Another advantage of this approach for data procurement is that it allows government agencies to become involved via scientific and regulatory oversight committees, which provides an important mechanism for guidance and peer review in the design and conduct of studies before the initiation of expensive data collection efforts. This mechanism also offers government agencies the opportunity to review data shortly after these are produced (providing more rapid access than awaiting the results of a published study), and it allows for important intermittent feedback and interpretation to facilitate revisions to study design and collection of subsequent definitive data sets. Although concern has been expressed regarding potential conflicts of interest associated with this approach for interaction of scientists from regulatory agencies and those representing the regulated community (Michaels and Wagner 2003), it has been used successfully in the U.S. for more than 15 years. Further, the resulting databases have been produced under Good Laboratory Practices (GLP; CFR 40 part 160), providing compliance with accepted standards of quality control and assurance.

Typically, when a task force is formed the data needs are established as part of an initial "scoping process." This process encourages conducting studies to fill identified data gaps. Task forces also continue to support the database they develop through a variety of adjunct activities such as the development of case studies, data management and modeling tools, and scientific publications. These adjunct activities serve to enhance and extend the useful lifetime of the database. Task force data development programs typically result in well-designed, -executed, and -documented databases that are more directly relevant to the products and exposure scenarios of scientific and regulatory interest. Further, these databases typically provide better data quality and quantity (number or replicate measures) in comparison to what are available in the peer-reviewed scientific literature. A key output of many task force programs is the development of a generic database that archives the raw data, provides useful summaries of the data and allows the data to be queried, subset, and analyzed for practical application in quantitative exposure assessments.

One of the driving factors in task force development is producing a large volume of data in a minimal time period. Generic databases are predicated on the magnitude of exposure, being generally independent of physico-chemical properties of the specific chemicals being studied. The basic premise of generic worker exposure databases, for example, is that mixer, loader, and applicator exposures can be estimated as a "normalized" function of the total amount of active ingredient (a.i.) handled, stratified by the type of application equipment, formulation, packaging, level of clothing protection, and specific worker tasks and practices. This generalized approach contrasts with reliance upon chemical-specific measurements for each active ingredient. Because of the larger pool of data that a generic database provides, there is the advantage of greater reliability in the resulting exposure estimates and a better understanding of the physical parameters and use conditions/work practices that affect exposure, which allows generalizations that are broadly applicable to a range of chemistries. The databases frequently support a set of equations or algorithms that describe complex exposures in terms of a single dependent variable (Table 1). They advance the science by providing for a means of hypothesis testing. Additionally, they provide the basis for definitive estimates of clothing/personal

Table 1. Overview of Generic North American Databases.

Task force	Generic data	Year formed	Independent variable	Dependent variable	Estimated number of studies[a]	Estimated number of replicates
PHED[b]	All handlers	1986	Pounds handled, time, none	mg/lb	60	1,700
ORETF[c]	Turf handlers/ reentry	1994	Pounds handled/ TTR,[g] time	mg/lb and cm^2/hr	10	200
ARTF[d]	Agricultural reentry	1995	DFR,[h] time, crop height	cm^2/hr	45	650
AHETF[e]	Agricultural handlers	2001	Several[i]	mg/lb	50	750
AEATF[f]	Antimicrobial handlers and reentry	2005	Several	mg/lb	19	270

[a]At this time there is no overalap of studies, i.e., no PHED studies are incorporated into any other database, etc.
[b]PHED, Pesticide Handlers Exposure Database.
[c]ORETF, Outdoor Residential Exposure Task Force.
[d]ARTF, Agricultural Reentry Exposure Task Force.
[e]AHETF, Agricultural Handlers Exposure Task Force.
[f]AEATF, Antimicrobial Exposure Assessment Task Force; proposed scope of ongoing program.
[g]TTR, transferable turf residue.
[h]DFR, dislodgeable foliar residue.
[i]Pounds handled, time, area treated, gallons sprayed, number of loads, or no normalization.

protective equipment (PPE)/engineering control protection factors. One of the most important aspects of a generic database is that it provides bounds to variability. Larger databases allow for robust distributional analyses and characterization of sample variance. Another opportunity provided by generic databases is the development of harmonized, consistent data interpretation. Finally, databases provide a benchmark for comparison to biomonitoring or other means of validation (Woollen 1993).

The identification of data gaps by government and industry and resulting government "data call-ins" are significant motivating factors in production of generic data (EPA 1994). Coordinating data development prevents "reinventing the wheel" so that the same type of data does not have to be produced independently by many different companies. This economy of scale is important because a typical "replicate" of data in a single study costs $10–20,000 U.S., and there are typically 15–20 replicates per study. The cost of generating new exposure data for the combinations and permutations of chemicals, formulations, equipment types, application rates, engineering control options, and PPE is prohibitive. Generic data are typically adequate for commonly used methods but may not cover unique or unusual practices. Because exposure-monitoring data are very expensive to produce or reproduce, task forces also typically offer some type of exclusive data rights. By different mechanisms the laws and regulations in both the U.S. and Canada protect access to the data developed by task forces. Permission or compensation is required for nonmember citation, but the entire database including the underlying raw data is freely available to government regulatory agencies.

Exposure studies residing in these databases have been conducted on a variety of pesticides, using commercial applicators, farmers, field workers, pest control operators (PCOs), professional lawn care operators, and homeowners engaged in normal activities. Several general approaches for quantifying pesticide exposures for these individuals have historically been used, including the use of (1) patch (e.g., gauze pad) dosimetry; (2) glove dosimeter/hand rinse techniques; (3) whole-body dosimetry, and (4) biomonitoring. The use of whole-body dosimeters, which are usually sectioned into standard body part areas (e.g., upper legs, lower legs) before extraction and analysis, prevents the need to extrapolate from a small patch size to the whole body part. As a result, the newer databases rely almost exclusively on whole-body dosimetry. Hand washes, patch dosimetry, or whole-body dosimeters are methods for quantifying the amount of pesticide that contacts the skin or clothing of an individual and provide a measure of dermal exposure.

Biomonitoring uses the amount of pesticide or metabolite detected in urine or other biological substrates of exposed individuals along with pharmacokinetics and metabolism data to obtain an accurate estimate of the total amount of pesticide absorbed by the person via all routes (inhalation, dermal, and incidental oral ingestion); biomonitoring does not account for individual exposure pathways. As a result, biomonitoring is normally used as a validation tool instead of being part of generic databases. Use of patch or whole-body dosimetry helps pinpoint exposure patterns and body part-specific contributions to total exposure; this allows risk managers to evaluate PPE and equipment that can mitigate exposures. Personal air monitoring devices have been used to characterize inhalation exposures. A measured volume of air is collected in the breathing zone of the worker and analyzed to quantify the pesticide of interest.

Quantifying worker exposure to pesticide active ingredients is an integral part of risk assessments associated with the pesticide registration process in the U.S., Canada, Australia, and the European Union. The risk assessment process enables regulatory agencies and the agrochemical industry to predict the possibility of adverse human health effects attributed to the use of a specific pesticide used under specific label conditions. Estimating risk requires knowledge of both exposure and toxicity. Therefore, exposures to the active ingredient associated with a given pesticide formulation must be known with some degree of certainty.

III. Generic Exposure Factors Require Generic Exposure Monitoring Methods

There are generic factors, either deterministic estimates or distributions, that can be derived from exposure databases. Standard generic methods are required to produce the dependent variable data that allow the uniform application of the generic exposure equation to the chemical-specific data (see Table 1). For example, in the Agricultural Reentry Exposure Task Force (ARTF) there are transfer coefficients for a specific reentry scenario derived from measured exposure and measured dislodgeable foliar residue (DFR). DFR dissipation is compound specific and is dependent on application rate, dissipation rate, formulation, and leaf type, although the method of measuring the DFR is generic. The initial (time zero) deposition is generic, not chemical specific. The use of DFR data generated in one crop but used in another makes the data generic. Exposure is scenario specific, but measuring exposures also involves standard (generic) protocols. Measuring the exposure and DFR concurrently by the standard generic methods can generate a generic transfer coefficient (TC). For example, if the DFR is measured assuming single-sided leaf exposure, the resulting TC when corrected by a factor of 2 is directly analogous to a TC measured with double-sided DFR. Similarly if exposure is measured on the outside of work clothing, the TC may not be generic for dermal exposure without application of a factor to relate outer to inner dosimeter measurements.

IV. Evolution of Exposure Databases
A. Handler Exposure

Subdivision U (EPA 1984) contains a concise historical description of pesticide exposure monitoring. In the early days, the driving force for exposure monitoring was understanding the causation of pesticide exposure that resulted in illness. Initially inhalation exposure was examined (Griffiths et al. 1951), reflecting industrial hygienists historical concern with inhalation exposure. Shortly thereafter, the first dermal dosimetry was conducted (Batchelor and Walker 1954). These two key studies occurred soon after the first commercialization of organophosphate insecticides, the cause of many new worker illnesses at that time, and just before the introduction of the gas chromatograph, the primary analytical tool that reduced limits of detection by orders of magnitude. Subsequently, whole-body dosimetry (Miller et al. 1980) and fluorescence tracer monitoring (Franklin et al. 1981) followed refinements of patch dosimetry, described by Durham and Wolfe (1962). Biomonitoring coevolved with passive dosimetry of parathion exposure (Lieben et al. 1953; Durham and Wolfe 1962).

Development of a generic exposure database requires consistent monitoring techniques or a quantitative method of relating exposure from one set of studies to another. The first proposals to create generic exposure databases were for handlers and were detailed in a series of papers from a 1984 American Chemical Society Symposium (Reinert and Severn 1985; Hackathorn and Eberhart 1985; Honeycutt 1985). One of the first attempts to quantitatively relate handler exposure to amount of active ingredient handled is described by Franklin et al. (1981). A joint EPA–industry task force was formed, headed initially by Joe Reinert of the EPA. This task force and cooperation from the National Agricultural Chemical Association (NACA), now known as CropLife America, Pest Management Regulatory Agency of Health Canada (PMRA), and the U.S. Environmental Protection Agency (EPA) led to the creation of the Pesticide Handlers Exposure Database (PHED). A predecessor of the PHED was a small-scale database of handlers that was utilized by regulators between 1985 and 1992. Two notable uses of this database were the alachlor PD4 (EPA 1987) and the dinoseb emergency suspension (EPA 1986a).

As with other applications of the risk assessment process, a tiered approach for estimating exposures is often taken. The first, or screening-level, tier of the risk assessment process for worker exposures to pesticides frequently involves use of generic exposure data (Krieger et al. 1992; Whitmyre et al. 2001). These data may come from databases such as PHED, used in North America (EPA 1997a), the Predictive Operator Exposure Database (POEM) developed in the U.K. (Martin 1986), and the German model. PHED provides actual measured dermal and inhalation exposure data that can be retrieved for specific subsetting conditions (e.g., open mixing, open cab ground boom application). POEM provides an estimate of exposure derived from a mathematical model developed from actual exposure monitoring data that are based on the type of formulation, application method, application rate, etc.

More recently, the Agricultural Handlers Exposure Task Force (AHETF) initiated the Agricultural Handlers Exposure Assessment Database (AHED) (Johnson 2005). AHED is composed of only Good Laboratory Practice (GLP) studies, with stringent guidelines for analytical recovery, analytical variability, and measurement of whole-body exposures that address many of the limitations inherent in PHED. These limitations included that virtually none of the data were GLP; >70% of the replicates were nondetects; minimum detection limit ranged over five orders of magnitude; statistics beyond central tendency were meaningless (van Hemmen 1992); many replicates were of such short duration or handled so little material that they were not generally applicable (i.e., not generic); vapor pressure, a key determinant of clothing penetration (Thongsinthusak et al. 1991), was not available; and protection provided by PPE or engineering controls was frequently severely underestimated.

Once normalized exposure data are obtained from the PHED or AHED output reports, the exposure can be calculated. In the case of the exposure data being normalized based on pounds a.i. handled, daily exposures (E_d) can be calculated as follows:

$$E_d \text{ (mg/kg/day)} = \text{(mg/lb a.i.)} * \text{(lbs a.i./acre)} * \text{(acres/day)} / \text{(bw, kg)} \quad (1)$$

Currently, EPA uses their surrogate exposure tables, which summarize PHED by use scenario in a convenient tabular format (see Keigwin 1998), for all their Registration Eligibility Decision documents.

B. Reentry Worker Exposure

Gunther and Blinn (1955) recognized that foliar residues of pesticides declined with time. This observation led to the concept that excessive exposure and resulting illness could be prevented by not allowing the workers to reenter treated fields until residues had dissipated to "safe" levels. Reentry worker exposure monitoring studies were first performed as industrial hygiene measurements to understand relative exposure by route, and from this early work the predominant exposure route seemed to be dermal (Milby et al. 1964). These studies proved useful in determining conditions responsible for worker illness following reentry to some treated orchards (Popendorf and Spear 1974).

During the late 1960s and early 1970s, researchers found that worker reentry exposures declined with declining DFR. On this basis, EPA proposed initial guidelines for developing restricted entry intervals (REI) (EPA 1984). At approximately the same time, researchers derived an empirical measure of residue transferability known today as transfer coefficient (TC). Popendorf and Leffingwell (1982) observed that the TC differed by activity and type of crop. Zweig et al. (1985) and Nigg et al. (1984) demonstrated the generic nature of the TC concept for both row crops and orchard crops. Subsequently, it was shown that reentry exposure was related to the degree of body immersion in treated foliage (Krieger et al. 1990; van Hemmen et al. 1995; EPA 2000). An early attempt to relate worker productivity to exposure can be found in Spencer et al. (1995).

Shortly after the first REIs were promulgated in California in 1971, efforts were made to estimate REI more quantitatively. In the 1970s, several investigators used the occurrence of cholinesterase (ChE) inhibition as an index of harvester exposures to organophosphates (OPs; Worgan and Rozario 1995). Serat et al. (1975) estimated a REI for an OP insecticide based on the dissipation rate of DFR and the rate of inhibition of plasma ChE in peach harvesters exposed to the foliar residues. He proposed that the REIs should be set so that workers exposed each day to DFR would not experience >30% inhibition of plasma ChE. Although inhibition of plasma ChE can be an indicator of exposure, it is not as sensitive an indicator as urinary metabolite excretion or passive dosimetry. Also, plasma ChE inhibition may be much less sensitive to inhibition than red blood cell (RBC) or brain ChE as observed with some inhibitors.

Knaak (1980) published a method for determining REI for ChE inhibitors. This method relied on three published exposure monitoring studies in which peach or citrus harvesters (Richards et al. 1978; Popendorf et al. 1979; Spear et al. 1977) exposed to foliage treated with azinphos-methyl, phosalone, or parathion were monitored for ChE inhibition. The exposure periods monitored were of relatively short duration (5–10 d) and DFR were not maximal, but the RBC ChE levels were determined to have not been significantly inhibited. Knaak et al. used these three monitoring studies as the basis for "safe levels" of foliar DFR. By dermally dosing rats with varying amounts of ChE inhibitors, Knaak et al. established the relative dermal potency of the insecticides. This rat potency was then related to DFR "safe levels" using proportional equations so that the dermally applied ChE inhibiting potency of a reference compound could be adjusted to reflect the dermal potency of a compound whose "safe level" was being estimated.

There are several disadvantages to this method of determining REI, and they relate directly to the interpretation of generic data. First, the method was limited to ChE inhibitors. Second, it was not possible to determine the carry-over effect of multiple dermal exposures because the potency data were based on single-dose studies in rats. When humans were exposed for longer periods to OP insecticides during harvest, more significant ChE inhibition did occur (McCurdy et al. 1994). Finally, REIs based on such measures should be crop specific. When reentry workers were engaged in higher-exposure work tasks such as commonly occur in grapes, frank cases of poisoning have been observed to occur at the "safe level" established in another crop (O'Malley et al. 1991). Thus, it is critically important that generic data be broadly applicable but not overgeneralized.

The DFR, by analogy to other sorts of environmental exposures, is equivalent to a source strength term. Iwata et al. (1977) described the original method, although minor adaptations have been made by other investigators and discussed by Dong et al. (1992), ranging from leaf punch to whole-leaf sampling. A recent review of factors affecting dissipation curves and resulting curve fits can be found in Whitmyre et al. (2003). Worker exposures may also be calculated for any day postapplication, using DFR values obtained from the regression curve(s) and appropriate TC values.

The dermal TC is related to the degree of contact between crop and worker and is dependent upon crop height and density and the frequency and type of contact for specific work activities (e.g., weeding, pruning, cutting, sorting/bundling, harvesting). The basis for calculating worker exposure to treated foliage is shown in Eq. 2, where DFR is the estimated value on the day of reentry:

$$\text{Exposure } (\mu g/\text{day}) = \text{DFR } (\mu g/cm^2) \times \text{TC } (cm^2/\text{hr})$$
$$\times \text{task duration } (\text{hr/day}) \tag{2}$$

Popendorf and Leffingwell (1982) found a proportional relationship between DFR level and dermal exposure to organophosphates over a narrow range of values. The range of TCs observed by these authors was 800–61,000 cm^2/hr, depending in part on the crop type and work activity. Zweig et al. (1985) reported an average TC of approximately 5,000 cm^2/hr across the studies they examined based on one-sided DFR values. This empirical factor of 5,000 cm^2/hr has been proposed as a generic default for the estimation of dermal exposure when the DFR is known without the benefit of actual measurements of human exposure (Zweig et al. 1985; Nigg et al. 1984). The ARTF has conducted a "cluster" analysis to group together TC values for different crop types and work activity types that are anticipated to have similar exposure potential and similar exposure patterns (i.e., similar distributions of exposure across different body regions).

A complete public database on worker reentry exposures comparable to PHED for handlers does not exist at this time, although a summary of TCs as a function of work task and crop type gleaned from the ARTF studies has been compiled by EPA (EPA 2000). The database design concept and application of its information have been described by others (Nigg et al. 1984; Honeycutt 1985; Zweig et al. 1985). ARTF, consisting mainly of companies who manufacture and/or distribute pesticides, has developed a worker reentry database for use by government regulators and its member companies. This database is based on actual field studies sponsored by ARTF or purchased by ARTF and contains generic TCs that are representative of specific crop types and worker activities. These generic TCs can then be used to estimate worker reentry exposures in conjunction with chemical- and crop-specific DFRs. This database allows subsetting of data by key variables that may affect exposure levels. This worker reentry exposure database contains data on dermal worker exposures, concurrently measured DFRs, site location, meteorological conditions, and other ancillary information (e.g., formulation type, method of application, reentry times, and conditions). The database provides output reports on TCs on a whole-body or body part-specific basis. The latter provides guidance for potential exposure mitigation methods.

C. Residential Exposure

Many of the regulatory changes over the past 20 years impacting residential use pesticides had their origins in California. The California Department of Food and Agriculture (CDFA) began formally collecting illness records in conjunction with alleged pesticide overexposure in 1982. Those data were used by CDFA and the California Department of Health Services (CDHS) as a tool to raise awareness of particular chemicals and/or use scenarios associated with possible illness. Concern by CDHS and CDFA about residential illnesses related to pesticide use in or around the house prompted Bayer Corp. (current name is Bayer CropScience; previously Mobay and Miles Corp.) to submit an estimate of indoor exposure following broadcast use of propoxur on carpet (Hackathorn and Eberhart 1983). This was a screening level assessment, and the first of its kind with extremely conservative assumptions about dermal pesticide transfer and hand-to-mouth transfer. The scope of that report format was expanded to include dichlorodivinylphosphate (DDVP) and chlorpyrifos and was submitted by CDHS in 1985, under sole authorship of California regulators, to the pesticide regulatory authority in California, CDFA, at the time. The submitted report also contained a request to reevaluate the safety of all indoor uses of the three insecticides. The same report was ultimately published as symposia proceedings in the open literature (Berteau et al. 1989) and has become one of the most cited references regarding children's alleged overexposure to pesticides. The report concluded that broadcast use of all three pesticides produced absorbed dosages in young children in the range of milligrams active ingredient per kilogram body weight. CDFA issued a reevaluation notice in 1985 that required residential exposure monitoring data to be supplied by December 1989.

To better understand the true exposure potential of children contacting a treated surface, CDFA designed and conducted an exposure monitoring study using a choreographed regimen (Jazzercise) that was reproducible, involved whole-body exposure monitoring, and had a concurrent measurement technique to assess transferable residue: the California Roller. This study was conducted in 1988 and reported by Ross et al. (1990, 1991). Jim Vaccaro of Dow Chemical Inc. observed the initial CDFA Jazzercise study and subsequently conducted a biomonitoring study with human subjects using another choreographed routine that simulated children's activities on a treated surface. That study was published (Vaccaro et al. 1996) and demonstrated very similar exposure to the 20-min Jazzercise study (Evans 1999). A series of hypothetical exposure estimates based on surface wipes and/or air measurements were published (Naffziger et al. 1985; Fenske et al. 1990), and all suggested lower exposures than those in Berteau et al. (1989). While the initial concern for residential exposure was with indoor exposure to surface residues, that changed in the early 1990s and began to include turf (Black 1993; Eberhart and Ellisor 1993). EPA issued Standard

Operating Procedures in the late 1990s to introduce a uniform means of calculating residential exposure and used an indoor study of Jazzercise on carpet (Formoli 1996) to characterize both indoor and outdoor residential exposures (Evans 1999).

When the Outdoor Residential Exposure Task Force (ORETF) was formed in late 1994, government agencies (EPA, PMRA, and CDPR) had reached a mutual understanding with ORETF that industry would generate data for residential turf exposure that would relate transferable turf residue (TTR) to dermal exposure (DE). ORETF members had previously generated several Jazzercise studies that related DE to TTR measured using the original California Roller (Ross et al. 1991). Subsequently, ORETF decided to develop a uniform standard method, the Modified California Roller (MCR), that could be used by all ORETF members to generate comparable TTR data. In the process, ORETF examined virtually all the existing methods of measuring TTR (Klonne et al. 2001; Rosenheck et al. 2001). The modification from the original California Roller was adding a longer handle on the roller and fixing the percale sheet matrix in a frame. After validating the utility, reproducibility, and comparability of the MCR to other methods of measuring TTR, all ORETF member companies produced TTR data using the new MCR to satisfy EPA's Data Call-In. Additionally, in an independent, round-robin testing protocol, EPA's Office of Research and Development found that the MCR performed more reproducibly and was easier to use than the polyurethane foam (PUF) roller or the drag sled (Fortune 1997). Williams et al. (2003) demonstrated that the California Roller was quite "rugged" in that neither weight of the roller nor number of rolls tended to have a significant effect on TTR, which further suggested that data from the CR and MCR should be very similar. CDPR recently tested this hypothesis and found that the CR described by Bernard et al. (2001) and MCR produced statistically significant difference in transferable residues for liquid, but not granular, pesticide formulations (Welsh et al. 2005). However, from the perspective of the log-log relationship of exposure to TTR described in the following paragraph, a two- to threefold difference appears to have little influence on the estimated exposure.

The generality of the relationship between transferable residue (TR) and DE has never been tested over a wide range of residues. A number of individuals have developed data over a period of years that suggest DE is not a linear function of TR. Popendorf and Leffingwell (1982) appear to be the first to publish a quantitative relationship between DE and DFR during postapplication agricultural reentry activities. The units of their TC were $\mu g\text{-}cm^2/ng\text{-}hr$ (i.e., DFR/DE), but graphical presentation of their data suggests the relationship was a power function as the data were plotted on a log scale on both axes. Popendorf and Leffingwell (1982) showed a series of parallel lines describing log dermal exposure as a function of log DFR, depending on the work function being measured. Durkin et al. (1995) observed a statistically significant linear regression of log hand DE versus log DFR. The International Life Sciences Institute (ILSI 2004) has compiled a summary of residential exposure monitoring studies and derived a relationship between log TR and log DE.

V. Conservative Exposure Estimates from Generic Data

By design, some generic databases (e.g., PHED, ORETF) tend to be "conservative," i.e., they tend to overestimate exposure. Furthermore, first-tier assessments using the databases tend to incorporate factors that compound that conservatism. For example, the best fitting curve describing the relationship of log DE as a function of log TTR is extremely conservative (ILSI, 2004); this exposure factor alone produces an upper-bound estimate. Typically, several conservative factors go into estimating exposures using the proposed algorithms. In the case of dermal estimates for residents exposures are estimated for naked individuals continuously engaged in high-contact aerobic exercise. Additionally, for residential reentry there are further factors that compound this conservative estimate. Among these factors are (1) the use of upper-bound TTR values (i.e., if TTRs are taken in multiple geographic locations, only the "average" values from the location producing the highest TTR will likely be used) (Smegal et al. 2001); (2) time on turf is assumed to be 2 hr, representing an upper-bound 75th percentile estimate; (3) exposures are estimated at the earliest reentry time allowed; and (4) the degree of contact was more intensive and frequent than children videotaped during normal play. Thus, the resulting deterministic estimate using these conservative input values will meet or exceed the theoretical upper-bound estimate (TUBE) of exposure for individuals reentering treated turf. This assertion is supported by biomonitoring under normal conditions showing that residential dermal exposure calculated using transferable residue are significantly higher (up to two orders of magnitude) than those estimated from urinary biomonitoring results (ILSI 2004). Although the available biological monitoring data have limitations (e.g., number of replicates, scenarios addressed), they suggest that potential conservative biases exist with predictive exposure estimation methods and indicate the need for further research and comparative analyses to support evaluation and validation of predictive exposure assessment methods.

The use of a plausible upper-bound estimate of exposure is consistent with EPA policy:

> "The first step that experienced assessors usually take in evaluating the scenario involves making bounding estimates for individual exposure pathways. The purpose of this is to eliminate further work on refining estimates for pathways that are clearly not important. The method used for bound-

ing estimates is to postulate a set of values for the parameters in the exposure or dose equation that will result in an exposure or dose higher than any exposure or dose expected to occur in the actual population. The estimate of exposure or dose calculated by this method is clearly outside of (and higher than) the distribution of actual exposures or doses. If the value of this bounding estimate is not significant, the pathway can be eliminated from further refinement. The theoretical upper bounding estimate (TUBE) is a type of bounding estimate that can be easily calculated and is designed to estimate exposure, dose, and risk levels that are expected to exceed the levels experienced by all individuals in the actual distribution. The TUBE is calculated by assuming limits for all variables used to calculate exposure and dose, that, when combined, will result in the mathematically highest exposure or dose. It is not necessary to go to the formality of the TUBE to assure that the exposure or dose calculated is above the actual distribution, however, since any combination that results in a value clearly higher than the actual distributions can serve as a suitable upper bound." (EPA 1992)

EPA has pointed out (Evans 1999) that the procedures outlined in the original Standard Operating Procedures for Residential Exposure Assessment (an example of screening-level exposure assessment methodology used to estimate upper-bound values) were also *not meant to be aggregated* without a definitive characterization by the assessor because it violates the basic tenets of exposure assessment by adding highly conservative estimates of exposures that result in "bounding, unrealistic estimates of exposure."

Since the 1990s there has been increasing interest in protecting the more highly exposed individuals from adverse toxicological effects potentially occurring after a single exposure. Terms such as "worst case," "maximally exposed individual," "extreme case," and "realistic upper bound" have been used to describe this exposure. It is the regulator's position that it is imperative that exposure is not underestimated, and in reality they have made substantial progress in moving away from defaults that were clearly overestimates to using generic data that tend to be somewhat greater than the central tendency of the distribution. In some cases the estimates are not internally consistent, e.g., high surface area and low body weight, or light work respiration rate with sedentary work conditions. In other cases, despite years of collecting use data, the maximum label rate rather than the actual use rates are employed. This tendency to use multiple upper-bound exposure inputs from generic databases produces a multiplicative effect.

Burmaster and Bloomfield (1996), Bogen (1994), Cullen (1994), and Burmaster and Harris (1993) more clearly define the principle of compounding conservatism. EPA's Scientific Advisory Panel (SAP) summarized the conclusion well with the following quote. "When inflated 'central tendency' values are put into the deterministic exposure calculation, they can be expected to overestimate the expected or 'central tendency' exposure. If the distribution of exposure is highly positively skewed, this bias may be considerable. In some cases the arithmetic mean values are substantially skewed and should be replaced by median values as a better indicator of central tendency. Working with high end values will be even worse, as the result will correspond to the very rare event of an exposure that is extreme in every respect and hence will be higher than is ever observed in reality." (SAP 2001).

Enumerated below are descriptions of several variables that make use of generic exposure data an overestimate of actual exposure and provide an approximate numerical indicator of that overestimate.

A. Central Tendency Statistic

Exposure monitoring results are typically skewed log-normally based on a statistical test for normality, and the appropriate central tendency statistic should be the geometric mean. Although this concept has been incorporated into the exposure assessment practice of some regulatory agencies [USEPA (PMRA) 1997], it is not used universally.

B. Body Weight

The average adult body weight default (70 kg) commonly used in exposure assessment is based on U.S. national survey data, but when used in occupational assessments this value implies that an average adult is representative across males and females in any given worker exposure scenario. However, most participants in exposure monitoring studies going into PHED, AHETF, ORETF, and ARTF were male adults, and the appropriate body weight to use in conjunction with these data is the mean male body weight of 77 kg (EPA 1997a). Even 77 kg is likely an underestimate, because surveys of PCO body weight typically show them averaging 90 kg (Krieger 2002) whereas the ORETF group of applicators averaged 86 kg. The AHED studies and those comprising POEM show applicator body weight to average 85 kg.

Exposure data derived from males normalized to a male body weight are directly applicable to even a toxicology endpoint occurring only in females for the purposes of calculating margin of exposure (MOE). The reason for this is that the ratio of surface area to body weight across the entire distribution for both males and females is virtually identical, and once a dose has been normalized to body weight, the resulting dosage is valid in either sex. To utilize exposure data measured in one sex to characterize dosage in another by merely changing the body weight is not appropriate.

C. Body Surface Area

Although this physiologic factor does not appear in many risk assessments, it is critical because it is used in calculating exposure either directly or indirectly. As mentioned in the previous paragraph, body weight and surface area are related. In calculating exposure, it is not appropriate to use $2.12 \, m^2$

as a total body surface [this is the common default in PHED data summary tables (Keigwin 1998) and represents the 85th percentile male or 95th percentile female surface area] in conjunction with a body weight of 70 kg, which represents a 30th percentile male or a 70th percentile female. The resulting conservatism of using an overly large surface area cannot be physiologically reconciled with the very low body weight. A more reasonable deterministic estimate of body surface area for males is the 50th percentile value of $1.94\,m^2$ (EPA 1997a) with a corresponding weight of 77 kg. AHED is being designed with greater flexibility to calculate exposure using an algorithm that links the replicate's body weight to surface area if the subject's height was recorded.

D. Respiration Rate

The respiration rate in the risk assessment is critical in converting exposure monitoring measurements of $\mu g/m^3$ to μg exposure per pound applied for use in generic databases. Historically, a standard default, and the one used in PHED data summary tables (Keigwin 1998), has been 29 L/min (EPA 1986b). This is a respiration rate consistent with a "moderate" level of exertion and is seldom experienced by pesticide handlers. A more reasonable respiration rate consistent with pesticide handling is 9–14 L/min, corresponding to sedentary and light work, respectively (EPA 1997a), consistent with driving a spray rig, etc. The use of higher respiration rates should be reserved for moderate work, such as walking up and down a ladder while carrying a load, and not for 8- to 12-hr exposure intervals.

E. Dermal Absorption

Many exposure assessments, especially those involving aggregate exposure, rely upon an estimate of dermal absorption to estimate a systemic dose to compare with oral-dose toxicity studies. There are three typical sources of overestimating systemic dose associated with dermal penetration. First, the default dermal absorption in the absence of data is 100% (Zendzian 1994), which is roughly double the upper-bound measurement for most chemicals in rats (Donahue 1996). Second, humans generally absorb considerably less pesticide through the skin than do rats (Ross et al. 2000). Finally, percent dermal absorption is dose dependent, and the most exposed anatomic regions such as hands will typically have lower bioavailability due to higher dose density than the remainder of the body (Ross et al. 2001). A dermal absorption of 10% would be representative for most pesticides.

F. Transferable Residue (DFR and TTR)

Measures of transferability are typically conducted in representative areas throughout the country. However, first-tier analysis of exposure typically uses the value that represents the highest average from one of several locations. Further, the shortest allowable reentry time is selected for collecting the transferable residue, although reentry is frequently hours to days longer than the shortest REI. Because residues typically decline exponentially with time, using the earliest TR to calculate exposure for any toxicological endpoint that requires more than a 1-d exposure is extremely conservative.

G. Maximum Rate/Maximum Number of Applications

Frequently, maximum application rate and maximum area applied/day are used to estimate exposure. This combination is inconsistent for several reasons. (1) Maximum rate is expensive and is not routinely used for many crops. Operators typically use the minimum application rate to get results, and not the maximum rate. California's extensive records of full use reporting confirm this (CDPR 1999–2003). (2) During a multiday time frame, a given applicator will not consistently use the maximum rate. Handlers who consistently use pesticides are professionals because they are operators working for themselves or some other licensed application firm. These individuals are employed by various companies, which represent an average use rate for that particular chemical. (3) Because the work is for hire, it is unlikely that each day the maximum acreage will be treated due to varying field sizes, equipment maintenance, equipment transport from site to site, use of alternative pesticides, inclement weather, holidays, and illness.

H. Residential Exposure Duration

Default assumptions of residence time in contact with a treated area are typically skewed towards the upper bound. For example, time on turf is assumed to be 2 hr or the 75th percentile based on NHAPS (EPA 1999). Similarly, time spent on a treated carpet (i.e., soft floor surfaces) is assumed to be 8 hr (Smegal et al. 2001). Adding to the conservatism is the assumption that individuals on turf or carpet are making full-body contact at the intensity level associated with aerobic exercise.

A shorter duration monitoring interval scaled up to estimate an 8-hr day usually overestimates exposure. DPR acknowledged this in the publication by Spencer et al. (1995), where short-duration monitoring overestimated a full day's exposure by twofold or more. Thus, using less than a full day's monitoring to estimate a full day's exposure will be conservative (Ross et al. 2000).

I. Effect of Using Combined Upper-Bound Exposure Factors

When EPA generates a Registration Eligibility Decision document, they usually acknowledge that the exposure estimates are upper bound. Typically, however, no attempt is made to quantify just how conservative or how representative those values are. It is critical that risk managers have some

Table 2. Partial Summary of Conservative Factors Typically Applied to Estimated Exposures Derived from Generic Databases.

Variable	Tier I[a]	Realistic[b]	Overestimate
Body weight (kg^{-1})	60–70	86	1.2–1.4
Body SA (cm^2)	21,200	20,700[c]	1.0
Respiration rate (L/min)	29	9–14	2.1–3.2
Dermal absorption (%)	100	10[d]	10
Transferable residue (%)	5–20	0.01–12	1.7–500
Application rate (lb/ac)	X	0.5–1×	1–2
Acres treated (ac/d)	40–1,200	20–350	2–3.4
Residential exposure Time (hr)	2–8	0.1–2	4–20

[a]Exposure defaults sometimes used in a first-tier assessment.
[b]Exposure factors more physiologically or agronomically consistent with "normal."
[c]For an 86-kg male.
[d]Average human dermal absorption of 13 different pesticides (Ross et al. 2001).

concept of the magnitude of uncertainty inherent in the risk calculations. In toxicology studies, this magnitude is given standard, somewhat conservative uncertainty factors for inter- and intraspecies differences (Swartout et al. 1998). Dosage to test animals is typically known with a high degree of certainty, and where it is not, additional uncertainty factors are clearly specified. Such is not the case with human exposure estimates. Rather, the uncertainty is frequently described as a range of values, and the chosen value may be either default (policy), "weight of evidence," or "professional judgment." The resulting risk estimates have nebulous uncertainty and may be deceptive to risk managers and the public. Table 2 is a summary of several factors used in calculating exposure from generic databases and how those factors tend to be upper bound in the resulting assessments.

Not all these factors are used in every exposure assessment. However, these factors are multiplicative in estimating exposure (see Eqs. 1 and 2), and the resulting degree of exposure overestimation can be severalfold. One method to check the exposure estimates from a generic database is to compare calculated exposure to the exposure measured with biomonitoring. As discussed for residential reentry exposure, generic estimates are typically high by severalfold compared to actual measurements. With the newest databases such as AHED, it may be possible to quantify the overestimation by comparing the exposure range from replicates measured over a full day's work versus standard operating procedure (SOP) defaults.

A key factor in developing generic exposure estimates is the use of consistent anatomic and physiologic factors such as body weight, body surface area, and respiration rate. The EPA Exposure Factors Handbook (EPA 1997a) and companion Children's Exposure Factors Handbook (EPA 1997b) are essential elements in converting exposure monitoring data into dose estimates and are another example of generic databases. EPA's (1989) Risk Assessment Guidance for Superfund Assessments – Human Health Evaluation Manual and subsequent EPA (1992) Guidelines for Exposure Assessment have been key to more uniform use of generic exposure data.

VI. Conclusions

Governments have embraced quantitative risk assessment as the vehicle for evaluating and justifying regulations regarding worker and residential safety. Quantitative risk assessment relies on credible exposure estimates. Beginning in the 1990s, numerous task forces have been created to develop exposure data for a variety of chemicals and associated formulation types, application equipment, and worker practices. Exposure assessment is a rapidly advancing science, and several task forces have been created to develop generic databases that address the scientific and regulatory needs of industry and government. Clearly, developing chemical-specific data for every product use pattern and associated variations would be prohibitively expensive and time consuming. A practical and scientifically valid alternative approach can be based upon extrapolation from a core set of exposure monitoring data compiled in the form of generic databases. Starting in the 1950s, research to quantitatively assess pesticide exposure was driven by clear evidence of overexposure sometimes observed in workers. However, the concern has been increasingly driven by prospective risk assessments with the goal of preventing excessive exposures or ensuring that acceptable probabilities for adverse effects are not exceeded. Prospective risk analysis, in conjuction with more rigorous safety standards, have created a demand for more robust exposure measurements databases.

Generic exposure databases represent a cost-effective and scientifically credible approach for meeting this demand. These databases provide a large number of replicate measurements and provide an array of associated demographic and ancillary information to facilitate hypothesis testing and the development of generalized exposure assessment methods across diverse chemistries. Generic databases provide opportunity for analysis of variance, distributions, correlations, and exposure estimatation, while maintaining quality control and achieving cost-effectiveness. Generic databases are useful for screening-level analyses but also support higher-tier probabilistic assessments. Exposures estimated from generic databases, in conjunction with other commonly used input variables, may result in conservative biases; therefore, consideration must be given to predictive method validation (e.g., comparison to estimates derived from relevant biological monitoring data), and to overall evaluations of variability and uncertainty in the context of risk analyses.

Due to the strength of generic databases and their potential role in global business practices, it is important that international harmonization among regulatory agencies continues to be pursued. Over the past few years, regulatory agencies have been making a concerted effort to standardize exposure assessment methods (Smegal et al. 2001). Regulatory harmonization is a concept that has been popularized by the need of the Organization for Economic Cooperation and Development (OECD) and North American Free Trade Association members to produce common requirements to encourage free trade. The goal of these efforts is to harmonize the assumptions and data analysis for worker and residential exposure assessments so that pesticide exposure reviews and work can be shared among regulatory jurisdictions. Ongoing harmonization efforts have undoubtedly been encouraged by previous successful activities, including development of the draft OECD Occupational Exposure Assessment Guidelines and PHED.

Summary

Faced with the need to evaluate under what conditions chemicals can be used with "reasonable certainty of no harm" to workers and consumers, industry and government agencies have embraced quantitative risk analysis as a science-based approach for product development, regulatory evaluations, and associated risk management decision making. Beginning in the 1990s, a variety of industry-sponsored task forces have been formed to develop exposure-related data to support safety evaluations for pesticide chemicals used in agricultural, industrial, institutional, residential, and other settings. Human exposure assessment and the underlying data (e.g., personal exposure and biological monitoring measurements, media-specific residue measurements, product use, and time–activity information) represent a critical component of the risk assessment process and a rapidly advancing science. While task forces have been created to develop databases for supporting the continued safe use of products, the development of these databases has served to advance general understanding of the basic principles underlying exposure assessment methodology and thereby provide the basis for improved science-based risk management by both industry and government. Given that developing chemical-specific data for every product use pattern and associated worker or consumer exposure scenario (e.g., professional mixer, loader and applicator activities associated with the use of a low-pressure sprayer, consumer residential lawn application via a ready-to-use hose-end sprayer product) is prohibitively expensive and time consuming, alternative approaches have been developed based upon meta-analyses and generalizations derived from databases of exposure monitoring studies for multiple chemicals, sorted by significant exposure covariates such as formulation type, method of application, amount of active ingredient applied, site of application, protective equipment and clothing, and task or activity. These generalizations can be used for predictive exposure analyses and have clearly demonstrated the value of "generic databases." Although data in these databases and associated generalizations are subject to interpretation, e.g., during the regulatory decision-making processes, and may be used in conjunction with additional considerations or assessment methods that result in conservative biases, the role of generic databases for risk management decision making, and advancing the science of applied exposure analysis continues to be realized.

References

AIR (Agriculture and Agro-Industry including Fisheries) (1996) The development, maintenance, and dissemination of a European Predictive Operator Exposure Model (EUROPOEM) database. A EUROPOEM database and harmonised model for prediction of operator exposure to plant protection products. A concerted action under the AIR (Agriculture and Agro-Industry including Fisheries) specific programme of the Community's Third Framework Programme for Research and Technological Development, and managed by DGVI.FII.3. Draft Final Report, December 1996. Publ no. AIR3 CT93-1370.

Batchelor, GS, and Walker, KC (1954) Health hazards involved in the use of parathion in fruit orchards of North Central Washington. Am Med Assoc Arch Ind Hyg 10:522–529.

Bernard, CE, Nuygen, H, Truong, D, and Krieger, RI (2001) Environmental residues and biomonitoring estimates of human insecticide exposure from treated residential turf. Arch Environ Contam Toxicol 41:237–240.

Berteau, PE, Knaak, JB, Mengle, DC, and Schreider, JB (1989) Insecticide absorption from indoor surfaces: hazard assessment and regulatory requirements. In: Wang, RGM (ed) Biological Monitoring for Pesticide Exposure. American Chemical Society Symposium Series No. 382. ACS, Washington, DC.

Black, KG (1993) An assessment of children's exposure to chlorpyrifos from contact with a treated lawn. Doctoral dissertation, Rutgers University. UMI Dissertation Services #9333377, Ann Arbor, MI.

CDPR (1999–2003) California State Pesticide Use Report (www.cdpr.ca.gov/docs/pur), by chemical. Cal/EPA Department of Pesticide Regulation, Sacramento, CA.

CFR 40 part 160 Code of Federal Regulations (1989) Good Laboratory Practice Standards.

Donahue, J (1996) Revised policy on dermal absorption default for pesticides. HSM-96005. Worker Health and Safety Branch, California Department of Pesticide Regulation, Cal/EPA, Sacramento, CA.

Dong, MH, Krieger, RI, and Ross, JH (1992) Calculated reentry interval for table grape harvesters working in California vineyards treated with methomyl. Bull Environ Contam Toxicol 49:708–714.

Durham, WF, and Wolfe, HR (1962) Measurement of the Exposure of Workers to Pesticides. Bull WHO 26:75–91.

Durkin, PR, Rubin, L, Withey, J, and Meylan, W (1995) Methods of assessing dermal absorption with emphasis on uptake from contaminated vegetation. Toxicol Ind Health 11:63–79.

Eberhart, DC, and Ellisor, GK (1993) Evaluation of Potential Exposure Resulting from Contact with BAYLETON-Treated Turf: Bayer Report No. 105137. EPA MRID No. 43125401. EPA, Washington, DC.

EPA (1984) Pesticide assessment guidelines. Subdivision K – Exposure: Reentry protection. 540/9-84-001. EPA, Washington, DC.

EPA (1986a) Decision and Emergency Order Suspending the Registrations of All Pesticide Products Containing Dinoseb. FR 51 3364 (Oct 14, 1986).

EPA (1986b) Pesticide assessment guidelines. Subdivision U Applicator exposure monitoring. Report #540/9-87-127. U.S. Environmental Protection Agency, Washington, DC.

EPA (1987) Alachlor: Special Review Position Document 4, U.S.EPA Office of Pesticide Programs, Washington, DC.

EPA (1989) Risk Assessment Guidance for Superfund Assessments: Human Health Evaluation Manual. Report #540/1-89/002. Washington, DC. U.S. Environmental Protection Agency.

EPA (1992) Exposure Assessment Guidelines. U.S. Environmental Protection Agency Office of Research and Development. Fed Reg 57(104):22888–22938.

EPA (1994) EPA Pesticide Registration Notice 94-9. Announcing the Formation of Two Industry-Wide Task Forces: Agricultural Reentry Task Force and Outdoor Residential Exposure Task Force. EPA, Washington, DC.

EPA (1997a) Exposure Factors Handbook. EPA/600/P-95/002Fa,c. Office of Research and Development, EPA, Washington, DC.

EPA (1997b) Child-Specific Exposure Factors Handbook. http://cfpub.epa.gov/ncea/.

EPA (1999) Science Policy Document (Draft) (March 8, 1999). Exposure Data Requirements for Assessing Risks from Pesticide Exposure of Children. [In that document, SOP 2.2 is "Postapplication Dermal Potential Dose from Pesticide Residues on Turf."] Summary available at http://www.pestlaw.com/x/guide/1999/EPA-19990308B.html.

EPA (2000) Science Advisory Council for Exposure Policy Number 003.1 Regarding: Agricultural Transfer Coefficients, Revised August 7. U.S. Environmental Protection Agency, Office of Pesticide Programs, Washington, DC.

Evans, J (1999) Overview of issues related to the standard operating procedures for residential exposure assessment. Presented to the EPA science advisory panel for the meeting on September 21, 1999. EPA Washington, DC.

Fenske, RA, Black, KG, Elkner, KP, Lee, C, Methner, MM, and Soto, R (1990) Potential exposure and health risks of infants following indoor residential pesticide applications. Am J Public Health 80:689–693.

Formoli, TA (1996) Estimation of Exposure of Persons in California to Pesticide Products That Contain Propetamphos. HS-1731. California Environmental Protection Agency, Sacramento.

Fortune, CR (1997) Evaluation of methods for collecting dislodgeable foliar residues from turf. EPA/600/R-97/119. Human Exposure and Atmospheric Sciences Division RTP, EPA National Exposure Research Laboratory, Washington, DC.

Franklin, CA, Fenske, RA, Greenhalgh, R, Mathieu, L, Denley, HV, Leffingwell, JT, and Spear, RC (1981) Correlation of urinary pesticide metabolite excretion with estimated dermal contact in the course of occupational exposure to Guthion. J Toxicol Environ Health 7:715–731.

Griffiths, JT, Stearns, CR, and Thompson, WL (1951) Parathion hazards encountered spraying citrus in Florida. J Econ Entomol 44:160–163.

Gunther, FA, and Blinn, RC (1955) Analysis of Insecticides and Acaricides. Wiley Interscience, New York.

Hackathorn, D, and Eberhart, D (1983) Risk assessment: Baygon use for flea control on carpets. DPR registration document no. 50021:126. California Department of Pesticide Regulation, Sacramento, CA.

Hackathorn, DR, and Eberhart, DC (1985) Database proposal for use in predicting mixer-loader/applicator exposure. Am Chem Soc Symp Ser 273:341–355.

Honeycutt, RC (1985) The usefulness of farmworker exposure estimates based on generic data. Am Chem Soc Symp Ser 273:369–375.

ILSI (2003) Probabilistic exposure assessment meeting in Brussels, October, 2003.

ILSI (2004) HESI Residential Exposure Factors Database Users Guide. ILSI, Health and Environmental Sciences Institute, Washington, DC.

Iwata, Y, Knaak, JB, Spear, RC, and Foster, RJ (1977) Worker reentry into pesticide-treated crops. I. Procedure for the determination of dislodgeable pesticide residues on foliage. Bull Environ Contam Toxicol 18:649–655.

Johnson, D (2005) See www.exposuretf.com for background on formation of the Agricultural Handlers Exposure Task Force.

Keigwin, TL (1998) PHED Surrogate Exposure Guide: Estimates of Worker Exposure From the Pesticide Handler Exposure Database Version 1.1. EPA, Office of Pesticide Programs, Washington, DC.

Klonne, D, Cowell, J, Mueth, M, Eberhart, D, Rosenheck, L, Ross, J, and Worgan, J (2001) Comparative study of five transferable turf residue methods. Bull Environ Contam Toxicol 67:771–779.

Knaak, JB (1980) Minimizing occupational exposure to pesticides: techniques for establishing safe levels of foliar residues. Residue Rev 75:81–96.

Krieger, RI (2002) Survey of body weights of pest control operators in California, n = 212. Personal communication, Feb. 18, 2002.

Krieger, RI, Blewett, C, Edmiston, S, Fong, HR, Meinders, DD, O'Connell, LP, Schneider, F, Spencer, J, Thongsinthusak, T, and Ross, JH (1990) Gauging pesticide exposure of handlers (mixer/loader/applicators) and harvesters in California agriculture. Med Lavoro 81:474–479.

Krieger RI, Ross JH, and Thongsinthusak T (1992) Assessing human exposures to pesticides. Rev Environ Contam Toxicol 128:1–15.

Kromhout, H, and Vermeulen, R (2001) Temporal, personal and spatial variability in dermal exposure. Ann Occup Hyg 45:257–273.

Lieben, JR, Waldman, K, and Krause, L (1953) Urinary excretion of paranitrophenol following exposure to parathion. Ind Hyg Occup Med 7:93–98.

Lunchick, C (1988) A summary of surrogate worker exposure data prepared in anticipation of dinoseb cancellation hearings. Health Effects Division, Office of Pesticide Programs, USEPA, Washington, DC.

Maddy, K, Edmiston, S, and Richmond, D (1990) Illnesses, injuries and deaths from pesticide exposures in California 1949–1988. Rev Environ Contam Toxicol 114:57–123.

Martin, AD (1986) Estimation of exposure and absorption of pesticides by spray operators. Scientific subcommittee on pesticides and British Agrochemical Association joint medical panel. Paper PS 4221/SC 8001.

McCurdy, SA, Schneider FA, Steenland, K, Wilson, B, Krieger, R, Hernandez, B, Spencer, J, and Margetich, S (1994) Assessment of azinphosmethyl exposure in California peach harvester. Arch Environ Health 49:289–296.

Michaels, D, and Wagner, W (2003) Disclosure in regulatory science. Science 302:2073.

Milby, TH, Ottoboni, F, and Mitchell, HW (1964) Parathion residue poisoning among orchard workers. JAMA 189:351–356.

Miller, CS, Hoover, WL, and Culver, WH (1980) Exposure of pesticide applicators to arsenic acid. Arch Environ Contam Toxicol 9:281–288.

Naffziger, DH, Sprenkel, RJ, and Metzler, MP (1985) Down to Earth 41:7–10.

NAS (National Academy of Sciences) National Research Council (1983) Risk assessment in the federal government: managing the process. National Academy Press, Washington, DC.

Nigg, HN, Stamper, JH, and Queen, RM (1984) The development and use of a universal model to predict tree crop harvester pesticide exposure. Am Ind Hyg Assoc J 45:182–186.

O'Malley, M, and McCurdy, S (1991) Subacute poisoning with phosalone, an organophosphate insecticide. West J Med 153:619–624.

O'Malley, MA, Smith, C, O'Connell, L, Ibarra, M, Acosta, I, Margetich, S, and Krieger, RI (1991) Illness among grape girdlers associated with dermal exposure to methomyl. Worker Health and Safety Branch Report HS-1604. California Department of Food and Agriculture, Sacramento, CA.

Popendorf, WJ, and Leffingwell, JT (1982) Regulating OP pesticide residues for farmworker protection. Residue Rev 82:125–201.

Popendorf, WJ, and Spear, RC (1974) Preliminary survey of factors affecting the exposure of harvesters to pesticide residues. Am Ind Hyg Assoc J 35:374–380.

Popendorf, WJ, Spear, RC, Leffingwell, JT, Yager, J, and Kahn, E (1979) Harvester exposures to Zolone (Phosalone) residues in peach orchards. J Occup Med 21:189–194.

Reinert, JC, and Severn, DJ (1985) Dermal exposure to pesticides EPA's viewpoint. Am Chem Soc Symp Ser 273:357–368.

Richards, DM, Kraus, JF, Kurtz, P, Borhani, NO, Mull, R, Winterlin, W, and Kilgore, WW (1978) Controlled field trial of physiological responses to organophosphate residues in farm workers. J Environ Pathol Toxicol 2:493–512.

Rosenheck, L, Cowell, J, Mueth, M, Eberhart, D, Klonne, D, Norman, C, and Ross, J (2001) Determination of a standardized sampling technique for pesticide transferable turf residues. Bull Environ Contam Toxicol 67:780–786.

Ross, J, Thongsinthusak, T, Fong, HR, Margetich, S, and Krieger, R (1990) Measuring potential dermal transfer of surface pesticide residue generated from indoor fogger use. Chemosphere 20:349–360.

Ross, J, Fong, HR, Thonsinthusak, T, Margetich, S, and Krieger, R (1991) Measuring potential dermal transfer of surface pesticide residue generated from indoor fogger use: Using the CDFA roller method. Chemosphere 22:975–984.

Ross, JH, Dong, MH, and Krieger, RI (2000) Conservatism in pesticide exposure assessment. Regul Toxicol Pharmacol 31:53–58.

Ross, JH, Driver, JH, Cochran, RC, Thongsinthusak, T, and Krieger, RI (2001) Could pesticide toxicology studies be more relevant to occupational risk assessment? Ann Occup Hyg 45(suppl 1):5–17.

SAP (2001) EPA Scientific Advisory Report, Dec. 12, 2001.

Serat, WF, Mengle, DC, Anderson, HP, Kahn, E, and Bailey, JB (1975) On the estimation of worker reentry intervals into pesticide treated fields with and without exposure of human subjects. Bull Environ Contam Toxicol 13:506–512.

Smegal, D, Dawson, J, and Evans, J (2001) Recommended revisions to the standard operating procedures (SOPs) for residential exposure assessments. Science Advisory Council for Exposure, policy number 12. U.S. Environmental Protection Agency, Office of Pesticide Programs, Washington, DC.

Spear, RC, Popendorf, WJ, Leffingfwell, JT, Milby, TH, Davies, JE, and Spencer, WF (1977) Fieldworker's response to weathered residues of parathion. J Occup Med 19:406–410.

Spencer, J, Sanborn, J, Hernandez, B, Krieger, R, Margetich, S, and Schneider, F (1995) Long vs short monitoring intervals for peach harvesters exposed to foliar azinphos-methyl residues. Toxicol Lett 78:17–24.

Swartout, JC, Price, PS, Dourson, ML, Carlson-Lynch, HL, and Keenan, RE (1998) A probabilistic framework for the reference dose (probabilistic RfD). Risk Anal 18:271–282.

Thongsinthusak, T, Brodberg, R, Ross, JH, Gibbons, D, and Krieger, RI (1991) Reduction of pesticide exposure by using protective clothing and enclosed cabs. HS-1616. Worker Health and Safety Branch, California Department of Pesticide Regulation, Sacramento, CA.

US EPA (1997) U.S. Environmental Protection Agency Series 875. Occupational and residential exposure test guidelines, Group B: Postapplication exposure monitoring test guidelines. Version 5.3, July 24. Office of Prevention, Pesticides, and Toxic Substances, USEPA, Washington, DC.

Vaccaro, JR, Nolan, RJ, Murphy, PG, and Berbrich, DB (1996) The use of unique study design to estimate exposure of adults and children to surface and airborne chemicals. Std Tech Publ 1287. American Society for Testing and Materials, West Conshohocken, PA, pp 166–183.

van Hemmen, JJ (1992) Estimating worker exposure for pesticide registration. Rev Environ Contam Toxicol 128:43–54.

van Hemmen, JJ (1993) Predictive exposure modelling for pesticide registration purposes. Ann Occup Hyg 37:541–64.

van Hemmen, JJ, and van der Jagt, KE (2001) Innovative exposure assessment of pesticide uses for appropriate risk assessment. Introductory remarks. Ann Occup Hyg 45(suppl 1):S1–S3.

van Hemmen, JJ, van Goldstein Brouwers, YGC, and Brouwer, DH (1995) Pesticide exposure and re-entry in agriculture. In: Curry, PB, et al. (eds) Methods of Pesticide Exposure Assessment. Plenum Press, New York, pp 9–19.

Welsh, A, Powell, S, Spencer, J, Schneider, F, Hernandez, B, Beauvais, S, Fredrickkson, AS, and Edmiston, S (2005) Transfrable turf residue following imidacloprid application. HS-1860. Worker Health and Safety Branch, California Department of Pesticide Regulation, Sacramento, CA.

Whitmyre, GK, Ross, JH, and Lunchick, C (2001) Occupational exposure databases/models for pesticides. In: Krieger RI (ed) Handbook of Pesticide Toxicology. Academic Press, San Diego.

Whitmyre, GK, Ross, JH, Lunchick, C, Volger, B, and Singer, S (2003) Biphasic dissipation kinetics for dislodgeable foliar residues in estimating postapplication occupational exposures to endosulfan. Arch Environ Contam Toxicol 46(1):17–23.

Williams, RL, Bernard, CE, Oliver, MR, and Krieger, RI (2003) Transferable chlorpyrifos residue from turf grass and an empirical transfer coefficient for human exposure assessments. Bull Environ Contam Toxicol 70:644–651.

Woollen, BH (1993) Biological monitoring for pesticide absorption. Ann Occup Hyg 37:525–540.

Worgan, JP, and Rosario, S (1995) Pesticide exposure assessment: Past, present, and future. In: Curry, PB, et al. (eds) Methods of Pesticide Exposure Assessment. Plenum Press, New York, pp 1–8.

Zendzian, RP (1994) Dermal absorption of pesticides. Pesticide assessment guide-lines, Subdivision F, hazard evaluation, human and domestic animals. Series 85-3. Health Effects Division, EPA, Washington, DC.

Zweig, G, Leffingwell, JT, and Popendorf, WJ (1985) The relationship between dermal pesticide exposure by fruit harvesters and dislodgeable foliar residues. J Environ Sci Health B20(1):27–59.

Manuscript received June 2; accepted June 10, 2005.

Rev Environ Contam Toxicol 186:133–174 © Springer 2006

Ecotoxicological Evaluation of Perfluorooctanesulfonate (PFOS)

Susan A. Beach, John L. Newsted, Katie Coady, and John P. Giesy

Contents

I. Introduction

Perfluorinated alkyl acids (PFAAs) are synthetic, fully fluorinated, fatty acid analogues that are characterized by a perfluoroalkyl chain and a terminal sulfonate or carboxylate group (Fig. 1). The high-energy carbon–fluorine (C-F) bond renders these compounds resistant to hydrolysis, photolysis, microbial degradation, and metabolism by animals, which makes them environmentally persistent (Giesy and Kannan 2002). Perfluorinated compounds have been manufactured for more than 50 years and are

Communicated by George W. Ware.

S.A. Beach
3M Company, Environmental Laboratory, 935 Bush Avenue St. Paul, MN 55144, U.S.A.

J.L. Newsted
ENTRIX, Inc., 4295 Okemos Road, Suite 101, Okemos, MI 48864, U.S.A.

K. Coady
Department of Biology, Warner Southern College, Lake of Wales, FL 33859, U.S.A.

J.P. Giesy (✉)
Department of Zoology, National Food Safety and Toxicology Center, Michigan State University, East Lansing, MI 48824, U.S.A.

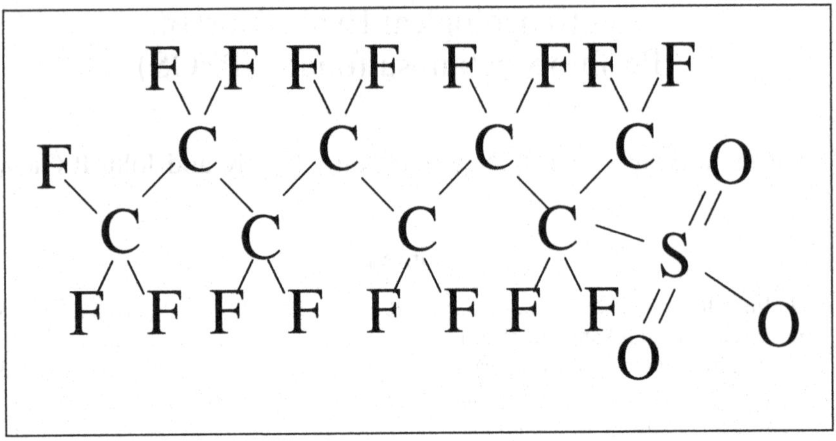

Fig. 1. Structure of perfluorooctane sulfonate (PFOS).

commonly used in materials such as wetting agents, lubricants, corrosion inhibitors, stain-resistant treatments for leather, paper, and clothing, and in foam fire extinguishers (Sohlenius et al. 1994; Giesy and Kannan 2002). The global environmental distribution, bioaccumulation, and biomagnification of several perfluoro compounds have been studied (Giesy and Kannan 2001).

Because PFAAs are chemically stabilized by strong covalent C-F bonds, they were historically considered to be metabolically inert and nontoxic (Sargent and Seffl 1970). Accumulating evidence has demonstrated that PFAAs are actually biologically active and can cause peroxisomal proliferation, increased activity of lipid and xenobiotic metabolizing enzymes, and alterations in other important biochemical processes in exposed organisms (Sohlenius et al. 1994; Obourn et al. 1997). In wildlife, the most widely distributed PFAA, perfluorooctane sulfonate (PFOS), accumulates primarily in the blood and in liver tissue (Giesy and Kannan 2001). PFOS appears to be the ultimate degradation product of a number of commercially used perfluorinated compounds, and the concentrations of PFOS found in wildlife are greater than other perfluorinated compounds (Kannan et al. 2001a,b; Giesy and Kannan 2002).

A large body of ecotoxicological information, generated over a period of more than 25 yr, exists for various salts of PFOS. However, until recently definitive information was not available on chemical purity, and validated analytical methodology did not exist to measure exposure concentrations in many of the early studies. Therefore, data generated before 1998 are somewhat unreliable relative to actual substance(s) used in the test as well as the fact that exposure concentrations were not measured as part of these studies. For this review, the potassium salt of PFOS was chosen by the 3M Company for laboratory study because it utilizes the most ecotoxicologically relevant cation of all the PFOS salts produced. The commercially prepared potassium product was available as a full-strength salt. Although lithium, ammonium, diethanolamine, and didecyldimethylammonium salts have been tested, many of these studies utilized mixtures containing 25%–35% active ingredients. In addition, the potassium salt of PFOS (PFOS-K$^+$) was produced in greater quantities than the other salts. For example, in 1997, PFOS-K$^+$ accounted for >45% of all PFOS salts produced (USEPA 2001). The primary ecotoxicological data used in this review are based on a series of studies utilizing a well-characterized sample of PFOS potassium salt. The majority of these studies were conducted in accordance with U.S. Environmental Protection Agency (USEPA) or Organization for Economic Cooperation and Development (OECD) Good Laboratory Practices. Older studies are also included in cases where more recently generated data were not available for various species. Further analysis of sample purity was conducted after the reporting of a number of the studies, resulting in adjustments to some reported concentration values that are reflected in subsequent documents but not in the original reports (USEPA Pollution Prevention and Toxics Docket: EPA AR226; oppt.ncic@epa.gov or http://www.epa.gov/epahome/dockets.htm). The corrected values are used in this review. In addition, this assessment also examines recent studies published in the open literature.

II. Environmental Fate and Transport
A. Physical/Chemical Properties

PFOS is moderately water soluble, nonvolatile, and thermally stable. The potassium salt of PFOS has a reported mean solubility of 589 mg/L in pure water. However, PFOS is a strong acid and in water at a neutral pH will dissociate into ionic components. Thus, the PFOS anion can form strong ion pairs with many cations, resulting in a salting-out effect in waters that contain great amounts of dissolved solids (Table 1). For instance, as salt content increases, the solubility of PFOS decreases, such that PFOS solubility in saltwater is approximately 12 mg PFOS/L as compared to 589 mg PFOS/L in pure water. PFOS has a reported mean solubility of 56 mg PFOS/L in pure octanol.

PFOS is not expected to volatilize based on its vapor pressure and predicted Henry's law constant. OECD (2002) classified PFOS as a type 2, involatile chemical that has a very low or possibly negligible volatility. Available physical/chemical properties for the potassium salt of PFOS are listed in Table 1.

B. Environmental Fate

Under environmental conditions, PFOS does not hydrolyze, photolyze, or biodegrade, and it is considered persistent in the environment.

Table 1. Physical/Chemical Properties of the Potassium Salt of Perfluorooctane Sulfonate (PFOS).

Parameter	Value	Reference
Melting point	$\geq 400°C$	Jacobs and Nixon 1999
Boiling point	Not calculable	OECD 2002
Specific gravity[a]	~0.6 (7–8)	OECD 2002
Vapor pressure	3.31×10^{-4} Pa @ 20°C	Van Hoven et al. 1999
Water solubility		
Pure water	680 mg/L	3M Company 2001e
	498 mg/L	van Hoven and Nixon 2000
Seawater	12.4 mg/L	3M Company 2001c
Octanol solubility	56 mg/L	3M Company 2001d
Log K_{ow}[b]	−1.08	OECD 2002
Henry's law constant[c]	$<2.0 \times 10^{-6}$	3M Company 2003

[a]pH values in parentheses.
[b]Log K_{ow} calculated with solubility of PFOS in water and n-octanol.
[c]Henry's law constant calculated at 20°C with the solubility for pure water.

Photolysis. To date, no evidence of direct or indirect photolysis of PFOS has been observed experimentally (3M Company 2001a). Using an iron oxide photoinitiator matrix model, the indirect photolytic half-life was estimated to be ≥ 3.7 yr at 25°C. This model was chosen because the experimental error was least in this matrix. This half-life is based on the analytical method of detection.

Hydrolysis. Under experimental conditions of 50°C and pH conditions of 1.5, 5, 7, 9, or 11, no hydrolytic loss of PFOS was observed in a 49-d study (3M Company 2001b). Based on mean values and precision measures, the half-life of PFOS was estimated to be ≥ 41 yr at 25°C. However, this estimate was influenced by the analytical limit of quantitation and that no loss of PFOS was detected in the study.

Biodegradation. Biodegradation studies where PFOS was monitored analytically for loss of parent compound have been conducted using a variety of microbial sources and exposure regimes (Gledhill and Markley 2000a–c; Lange 2001). In one study, no loss or biotransformation of PFOS was observed over a 20-wk period with activated sludge under aerobic conditions, nor were any losses observed in a study conducted for 56 d with activated sludge under anaerobic conditions. The findings from these studies are supported by the results from a Ministry of International Trade and Industry MITI-I test (Kurume Laboratory 2002) that showed no biodegradation of PFOS after 28 d as measured by net oxygen demand, loss of total organic carbon, or loss of parent material. In addition, no losses of PFOS were observed in a biodegradation study conducted with soil under aerobic conditions (Gledhill and Markley 2000b).

The results from these studies differ from those observed by Schroder and coworkers, who observed the disappearance of PFOS from wastewater under anaerobic conditions (Schroder 2003; Meesters and Schroder 2004). In these studies, PFOS was spiked into a bioreactor containing sludge under anaerobic conditions at approximately 5–10 mg PFOS/L. Aqueous PFOS concentrations were then monitored by flow injection analysis-mass spectrometry (FIA-MS) and by liquid chromatography (LC)-MS. After 2 d, PFOS concentrations in water decreased to below the detection limit. However, there was no evidence of mineralization in that PFOS degradation products were not measured in the water extracts nor were nonpolar, volatile fluorinated compounds detected in the digester gas. In addition, fluoride ion concentrations in water did not increase during the study, which would be indicative of mineralization. Although there was no adsorption of PFOS to the glass walls of the reactor, that did not account for potential losses from the water to suspended organic particles which were removed by filtration before analysis. In contrast, under aerobic conditions aqueous PFOS concentrations in the bioreactor did not decrease over time. These results would seem to indicate that adsorption of PFOS to organic material is influenced by the redox potential of the system such that under anaerobic conditions the adsorption increases over that observed under aerobic conditions. However, additional studies are necessary to better evaluate this phenomenon. As a result, the ecological relevance of these tests is difficult to interpret and, as of this date, no laboratory or field data exist that demonstrate that PFOS undergoes significant biodegradation under environmental conditions.

Thermal Stability.   Several studies suggest that PFOS would have relatively low thermal stability, which is based on the fact that the carbon–sulfur (C-S) bond energy is much weaker than carbon–carbon (C-C) or carbon–fluorine (C-F) bonds and, as a result, would break first under incineration conditions (Dixon 2001). This conclusion is confirmed by Yamada and Taylor (2003), indicating that PFOS should be almost completely destroyed in a high-temperature incineration system.

C. Partitioning

Octanol/Water Partitioning.   An octanol/water partition coefficient has not been directly measured for PFOS but has been estimated from its water and octanol solubilities. Other physiochemical properties such as bioconcentration factor and soil adsorption coefficient cannot be estimated with conventional quantitative structure activity relationship models (QSAR). The use of K_{ow} is not appropriate to predict these other properties, because PFOS does not partition into lipids but instead binds to certain proteins in animals (Jones et al. 2003). As a result, use of either water solubility or pre-

Table 2. Adsorption and Desorption of PFOS to Soils, Sediments, and Soils.[a]

Soil type	Adsorption kinetics		Desorption kinetics	
	K_d (L/kg)	$K_{ads}F^b$ (L/kg)	K_{des} (L/kg)	$K_{des}F^b$ (L/kg)
Clay	18.3	25.1	47.1	105
Clay loam	9.72	14.0	15.8	60.2
Sandy loam	35.3	28.2	34.9	94
River sediment	7.42	8.70	10.0	44.6
Domestic sludge	<0.120	338	<237	3,130

[a]Values of K_d and K_{des} are averaged values.
[b]Freundlich coefficient. The units for the isotherms assume that the term $n = 1$. More accurately, the units are $(\mu g^{1-1/n}(L)^{1/n} kg^{-1})$.

dicted K_{ow} values may underestimate the accumulation of PFOS into organisms and other environmental media.

Adsorption/Desorption. PFOS appears to adsorb strongly to soil, sediment, and sludge (Table 2) with a range of distribution coefficients (K_d) in soils between 9.7 L/kg (clay loam) and 35 L/kg (sandy loam) (3M Company 2001f). Once adsorbed to these matrices, PFOS does not readily desorb even when extracted with an organic solvent. The average desorption coefficient (K_{des}) for soil was determined to be less than 10 L/kg. Adsorption and desorption equilibrium in these matrices was achieved in less than 24 hr, with greater than 50% occurring after approximately 1 min of contact with the test adsorbents. As a result, PFOS exhibited little mobility in all matrices tested and would not be expected to migrate any significant distance. Because PFOS is a strong acid, it most likely forms strong bonds in soils, sediments, and sludge via a chemisorption mechanism.

Bioconcentration. In a study with the northern leopard frog (*Rana pipiens*), tadpoles were exposed to PFOS concentrations ranging from 0.03 to 10 mg PFOS/L for up to 56 d (Ankley et al. 2004). In general, whole-body PFOS concentrations at each exposure concentration increased by 30- to 100 fold by 10 to 20 d posthatch and attained a near steady-state condition by the end of the exposure. Accumulation of PFOS from water was also size dependent, with larger frogs accumulating PFOS to a greater degree than smaller frogs such that there was some overlap in whole-body concentrations between treatment groups. Using a one-compartment bioaccumulation model, growth and size were shown to have an impact on steady-state PFOS concentrations and on the uptake and elimination constants of PFOS. Overall, depending on size and exposure concentration, kinetically derived bioconcentration factor (BCF) (BCFK) values ranged from 27.7 to approximately 200.

The potential of PFOS to bioaccumulate into fish and the relative importance of dietary and water-borne sources of PFOS to fish have been evaluated. In a bioaccumulation study with juvenile rainbow trout (*Oncorhynchus mykiss*), fish were exposed to 0.54 µg/g PFOS in the diet for 34 d followed by a 41-d depuration phase (Martin et al. 2003a). PFOS was accumulated by and depurated from the liver and carcass in a time-dependent manner. The predicted time to reach 90% steady state would be 43 d, which was approximately the same exposure duration of the study. The liver and carcass depuration rate constants were 0.035/d and 0.054/d, representing depuration half-lives of 20 and 13 d, respectively. The assimilation efficiency was 120% ± 7.9%, which indicates efficient absorption of PFOS from ingested food relative to other compounds (Fisk et al. 1998). In addition, this assimilation efficiency of PFOS may be indicative of enterohepatic recirculation, which could affect the disposition of PFOS in fish. Evidence of enterohepatic recirculation in rats has been demonstrated to affect the rate of elimination (Johnson et al. 1984). The bioaccumulation factor (BAF) for PFOS was 0.32 ± 0.05, indicating that dietary exposure did not result in biomagnification in trout. This small BAF was most likely the result of several factors, including a relatively low experimental feeding rate (F = 1.5% body weight) coupled with a relatively rapid rate of depuration. Overall, these data show that, under these experimental conditions, the diet would not be a major route of PFOS exposure for fish.

Studies conducted with other fish species have shown that PFOS does bioconcentrate in tissues from waterborne exposures (Table 3). Bluegill exposed to 0.086 or 0.87 mg PFOS/L in a flow-through system accumulated PFOS into edible and nonedible (fins, head, and viscera) tissues in a time-dependent manner (Drottar et al. 2001). In this study, fish were exposed to 0.086 mg PFOS/L for 62 d but were only exposed to 0.87 mg PFOS/L for 35 d because of excessive mortality. At the end of the exposure phase of both treatments, PFOS concentrations in tissues appeared to still be increasing. As a result, kinetic analyses of the data were conducted to calculate the kinetic bioconcentration factor (BCFK) based on estimated uptake and depuration rate constants. Due to mortality of fish exposed to 0.87 mg PFOS/L, the data from this exposure were not used to estimate these parameters. The BCFK values for edible, nonedible, and whole-fish tissues were calculated to be 1,124, 4,013, and 2,796, respectively. During the elimination phase of the study, PFOS depurated slowly, and time to reach 50% clearance for edible, nonedible and whole fish tissues was 86, 116, and 112 d, respectively.

Tissue distribution and accumulation kinetics were determined in rainbow trout exposed to a mixture of 11 PFAAs in water that included PFOS (Martin et al. 2003b). In trout exposed to 0.35 µg/L PFOS, the greatest PFOS concentrations were measured in the blood > kidney > liver > gallbladder; the least PFOS concentrations were in the gonads > adipose > muscle tissue (see Table 3). In blood, approximately 94%–99% of the PFOS

Table 3. Kinetic Parameters and Bioconcentration Factors (BCF) of PFOS in Fish.

			Kinetic parameters			
Species	Tissue	Apparent BCF[a]	K_u (L/kg*d)	K_d (l/d)	BCFK[b] (L/kg)	Half-life (d)
Bluegill	Edible	484	9.02	0.0080	1,124	86
(86 µg/L)	Nonedible	1,124	24.1	0.0060	4,013	116
	Whole	856	17.4	0.0062	2,796	112
Rainbow trout	Carcass	–	53	0.048	1,100	15
(0.35 µg/L)	Blood	–	240	0.057	4,300	12
	Liver	–	260	0.050	5,400	14
Carp						
(2 µg/L)	Whole	1,300				152
	Liver	4,300				
(20 µg/L)	Whole	720				49
	Liver	2,100				

[a]Apparent BCF was calculated as the concentration in fish at the end of the exposure phase (assumed steady state) divided by the average water concentration.
[b]Kinetic parameters are as follows K_u is the uptake rate constant (L*kg^{-1}*d^{-1}), K_d is the depuration rate constant (d^{-1}); BCFK is the kinetic BCF estimated as K_u/K_d.

was associated with plasma and only a minor amount was associated with the cellular fraction. PFOS also accumulated in the gills, indicating their importance in the uptake and depuration in trout. In general, depuration rate constants determined for carcass, blood, and liver showed that PFOS was more rapidly depurated than some organochlorine contaminants (PCBs, toxaphene) but more slowly than that observed for other surfactants (Tolls and Sijm 1995; Fisk et al. 1998). When compared to other surfactants, the uptake rate constants were greater than expected and were directly related to greater tissue concentrations (Tolls et al. 1997). Bioconcentration factors calculated from kinetic rate constants (BCFK) were 1,100, 4,300, and 5,400 for carcass, blood, and liver, respectively. As was observed for bluegill, steady-state PFOS concentrations in tissues were not achieved at the end of the exposure period. The 12-d accumulation ratios (BCF divided by tissue concentration at the end of the exposure period) for carcass, blood, and liver were greater than 600, indicating that the tissues were far from steady state. However, the kinetic BCFs calculated for rainbow trout were well within the range of values observed for other species such as bluegill and carp.

In a flow-through bioconcentration study of PFOS conducted with carp (*Cyprinus carpio*), fish were exposed to 2 or 20 µg PFOS/L and water and fish tissue samples were collected throughout testing (Kurume Laboratory 2002). Upon sampling, fish were separated into parts including integument (skin except head, scales, fins, alimentary canal, or gills), head, viscera (internal organs except for alimentary canal and liver), liver, and carcass and analyzed for concentrations of PFOS. Kinetic analysis was not conducted with the uptake portion of the exposure period. Instead, BCFs were calculated in all fish tissues at steady state. Steady state was assumed when three or more consecutive sets of tissue PFOS concentrations were not statistically different. In fish exposed for 58 d, the apparent BCFs in carp from the 2-µg PFOS/L treatment ranged from 200 to 1,500. In fish from the 20-µg PFOS/L exposure, apparent BCFs ranged from 210 to 850. PFOS depurated slowly. The time to reach 50% clearance for fish in the 20-µg PFOS/L treatment was 49 d, whereas 152 d were required for fish in the 2-µg PFOS/L treatment to reach 50% clearance.

Laboratory studies have demonstrated that PFOS accumulates into fish and frogs in a time- and concentration-dependent manner. In addition, these studies suggest that the primary route of accumulation into fish is from aqueous exposure to PFOS and that dietary sources of PFOS are secondary and may not be an important contributing factor in terms of the overall accumulation of PFOS by fish. However, the importance of either pathway relative to the accumulation of PFOS into fish under environmental conditions is still uncertain. This conclusion is based on observed discrepancies between accumulation factors measured in the laboratory and those estimated from the field. For instance, BAFs calculated from liver and surface water PFOS concentrations ranged from 6,300 to 125,000 in the common shiner (*Notropus cornutus*) collected in a Canadian creek (Moody et al. 2001). In contrast, the BCF for rainbow trout based on liver concentrations was 5,400 (Martin et al. 2003b). The discrepancy between laboratory and field accumulation values has also been observed for fish collected from Tokyo Bay, Japan (Taniyasu et al. 2003). In that study, PFOS concentrations in fish livers resulted in BAFs that ranged from approximately 1,260 to 19,950. In a field study conducted in the Tennessee River near Decatur, Alabama, fish and surface-water samples were collected and analyzed for PFOS. BCFs based on surface-water PFOS concentrations and whole-body PFOS concentrations in catfish and largemouth bass ranged from 830 to 26,000 (Giesy and Newsted 2001), whereas laboratory whole-body BCFs have been shown to range from approximately 700 to 3,000 depending on species and exposure concentration.

While there is some agreement between BCFs derived from laboratory studies with different fish species, BAFs derived from field data varied greatly. In addition, field-based BAFs tended to be greater than laboratory BCFs by at least an order of magnitude. Potential contributors to the variability of field-based BAFs include variable and unquantified contributions of PFOS precursors that may be present in the environment. In addition, temporal and spatial variability in surface-water PFOS concentrations must be considered in evaluating the field-derived BAFs. For instance, in Moody et al. (2001), fish liver and surface-water PFOS concentrations used to estimate BAFs were collected 7 mon after the spill. Thus, if PFOS concentrations in surface waters dissipated at a greater rate than concentrations in fish depurated, the use of these data would result in an overestimate of the "real" BAF. The authors, however, assumed that the fish and water were in equilibrium, without data to substantiate this assumption.

In addition, field-based BAFs are affected by factors including species-specific differences in the accumulation and elimination of PFOS. Moreover, dietary sources of PFOS may be more important in the accumulation of PFOS by fish during their life cycle than would be expected based on results from laboratory studies conducted with rainbow trout. As a result, field-based BAFs incorporate a significant amount of uncertainty that makes their use in predicting fish tissue PFOS concentrations from surface waters problematic in that they most likely overestimate the actual accumulation process.

Few studies have been conducted to evaluate the bioaccumulation of PFOS into terrestrial plants or organisms. In a toxicity study with worms exposed to 77, 141, 289, 488, and 1,042 mg PFOS/kg soil dry weight (dw), concentrations of PFOS were measured in soil and worms after a 14-d exposure (Sinderman et al. 2002). The worms were allowed to clear soil from their guts before chemical analysis. To estimate the BAF, it was assumed that all the PFOS in soils was bioavailable and that the worms had reached a steady state by the end of the exposure period. In addition, no attempt was made to differentiate between uptake from soil moisture and the ingestion of soil by the worms. Because treatment-related effects in worms were noted at exposures equal to or greater than 141 mg PFOS/kg soil dw, only the 77-mg PFOS/kg soil dw (or 52 mg PFOS/kg soil ww) were used to estimate a BAF. Based on this approach and assumptions, the BAF for worms was 3.75.

Limited data are available for uptake of PFOS into terrestrial plants (Brignole et al. 2003). In one study, seven plant species were exposed to PFOS in the soil and plant tissues were analyzed at study termination (Table 4). The concentration of PFOS in plant vegetative tissues varied less than 10 fold across the seven species, with the greatest concentrations being observed in soybeans. The PFOS concentrations in plant vegetative tissues were generally about 1–2 times greater than that measured in the soil, with the greatest difference between soils and tissues being observed at the lowest soil concentrations. The greatest BAF was observed for soybeans where concentrations in the vegetative tissues were approximately 4 fold greater than soil PFOS concentrations. Concentrations in fruit were typically less than 10% of the soil level, except for onion. While the PFOS concentrations in the fruit of onions from the 3.6-mg PFOS/kg soil treatment were less than in soil, concentration in the fruit of the 11-mg PFOS/kg soil were approximately 2 fold that of soil. No general relationships were observed between fruit and vegetative tissues for any of the plant species.

Table 4. Bioaccumulation Factors (BAFs) Estimated for Seven Terrestrial Plants Exposed to PFOS in the Soil.[a]

		Soil concentrations (mg PFOS/kg, dw)[c]			
Species[b]	Tissue	3.61	11.1	50.8	278
Onion (67 d)	Vegetative	NR	0.95	–	–
	Fruit	0.87	2.0	–	–
Ryegrass (205 d)	Vegetative	2.3	2.8	0.96	0.24
Alfalfa (141 d)	Vegetative	1.7	0.38	0.22	0.06
	Fruit	NR	NR	NR	NR
Flax (94 d)	Vegetative	1.4	1.69	1.1	–
	Fruit	0.06	0.12	0.05	–
Lettuce (67 d)	Vegetative	2.4	0.95	0.83	–
Soybean (67 d)	Vegetative	4.3	3.2	1.2	0.41
	Fruit	0.39	0.08	0.02	0.01
Tomato (94 d)	Vegetative	NR	3.05	0.99	–
	Fruit	NR	0.09	0.04	–

[a]All tissue and soil data reported on a dry weight basis (dw).
[b]Numbers in parentheses indicate when samples were collected for each species.
[c]Soil concentrations based on day 0. NR, not reported, because tissue PFOS concentration was less than the limit of quantitation (LOQ), which varied.

III. Ecotoxicology
A. Aquatic

Activated Sludge Microorganisms and Bacteria. To evaluate the potential effects of PFOS on activated sludge, microbes from a municipal wastewater treatment plant were exposed to nominal concentrations ranging from 0.9 to 870 mg PFOS/L (Schafer and Flaggs 2000). After 3 hr of exposure, there was a 39% inhibition of the respiration rate at the greatest concentration when compared to controls. However, it should be noted that this test concentration is in excess of the water solubility for PFOS. Based on current environmental concentrations, PFOS would not be expected to cause any effects to wastewater treatment plant sludge communities (Table 5).

Phytoplankton. A number of studies have been conducted to determine the toxicity of PFOS to phytoplankton (Table 5). Because the life cycle of most aquatic microorganisms is short, ranging from hours to days, these studies represent the measurement of chronic effects on multiple generations even when the exposure period of these tests are short (3–5 d). The toxicological endpoints that have been evaluated in these studies include growth, measured in terms of cell density, and growth rate and/or area under the growth curve over the test duration. Reported EC_{50} values for effects on growth based on cell density ranged from 48 to 263 mg PFOS/L

Table 5. Acute (Short-Term) and Subchronic (Long-Term) Toxicity Data for Phytoplankton Exposed to PFOS.

Species	96-hr EC_{50} (mg/L)[a]	96-hr NOEC (mg/L)	Reference
Activated sludge			
Respiratory inhibition	>870	NR	Wildlife International 2000
Selenastrum capricornutum[b]			
Growth (cell density)	68 (63–70)	42	Drottar and
Inhibition of growth rate	121 (110–133)	42	Krueger 2000a
S. capricornutum			
Growth (cell density)	48.2 (45.2–51.1)	5.3 (4.6–6.8)	Boudreau et al. 2003a
Growth (chlorophyll a)	59.2 (50.9–67.4)	16.6 (8.5–28.1)	
Chlorella vulgaris			
Growth (cell density)	81.6 (69.6–98.6)	8.2 (6.4–13.0)	Boudreau et al. 2003a
Anabaena flos-aquae			
Growth (cell density)	131 (106–142)	93.8	Desjardins et al.
Inhibition of growth rate	176 (169–181)	93.8	2001a
Navicula pelliculosa			
Growth (cell density)	263 (217–299)	150	Sutherland and
Inhibition of growth rate	305 (295–316)	206	Krueger 2001
Skeletonema costatum			
Growth (cell density)	≥3.2	≥3.2	Desjardins et al. 2001b

[a]EC_{50} values with 95% confidence intervals in parentheses.
[b]The species *Selenastrum capricornutum* has been renamed *Pseudokirchneriella subcapitata*.

(Table 5). The 96-hr no observed effect concentration (NOEC) values based on total biomass ranged from 5.3 to 150 mg PFOS/L. Based on biomass data, the most sensitive species was *Selenastrum capricornutum* (NOEC = 5.3 mg PFOS/L), whereas the diatom *Navicula pelliculosa* was the least sensitive species (NOEC = 263 mg PFOS/L) (Sutherland and Krueger 2001; Boudreau et al. 2003a).

When growth rate was evaluated as the test endpoint, 96-hr EC_{50} values ranged from 121 to 305 mg PFOS/L while NOEC values ranged from 42 to 206 mg PFOS/L. Again *S. capricornutum* was the most sensitive species whereas *N. pelliculosa* was the least. The effects of PFOS on these algal species were algistatic because growth resumed when algae from the greatest PFOS exposure were placed in fresh growth media at the end of the exposure period. Furthermore, signs of aggregation or adherence of the

Table 6. Acute and chronic toxicity of PFOS to aquatic macrophytes.

Species	Test duration	EC$_{50}$ (mg/L)	NOEC (mg/L)	Reference
Lemna gibba (Frond number)	7 d	108 (46–144)	15	Desjardins et al. 2001c
Lemna gibba (Frond number)	7 d	59.1 (51.5–60)	29.2	
(Biomass)	7 d	31.1 (22.2–36.1)	6.6	Boudreau et al. 2003a
Myriophyllum spicatum (Biomass, dw)[b]	42 d	12.5 (6–18.9)	11.4	Hanson et al. 2005
(Root length, cm)	42 d	16.7 (10.8–22.5)	11.4	
Myriophyllum sibiricum (Biomass, dw)[b]	42 d	3.4 (1.6–5.3)	2.9	Hanson et al. 2005
(Root length, cm)	42 d	2.4 (0.5–4.2)	0.3	

[a]EC$_{50}$ values with 95% confidence intervals in parentheses.
[b]Biomass was measured as dry weight (dw) (g).

cells to the flask were not observed, nor were there any noticeable changes in cell morphology at the end of the studies for any concentration evaluated in these studies.

Although concentration–response relationships for growth have been developed for freshwater algae, the marine diatom *Skeletonema costatum* was not affected by exposure to PFOS. In this study, a 96-hr EC$_{50}$ could not be determined because growth was not significantly inhibited at the greatest dissolved concentration attained under test conditions (3.2 mg PFOS/L).

Aquatic Macrophytes. In a toxicity test with duckweed (*Lemna gibba*), the 7-d IC$_{50}$ based on the number of fronds produced was 108 mg PFOS/L (46–144 mg PFOS/L) and the NOEC was 15 mg PFOS/L (Table 6) (Desjardins et al. 2001c). Sublethal effects noted in *L. gibba* exposed to concentrations greater than or equal to 31.9 mg PFOS/L included root destruction and/or cupping of the plant downward on the water surface. At higher PFOS concentrations (147 and 230 mg PFOS/L), there was a concentration-dependent increase in dead, chlorotic, and necrotic fronds. In a second study with *L. gibba*, the 7-d IC$_{50}$ based on frond number was 59 mg PFOS/L (51.5–60.3 mg PFOS/L). The 7-d IC$_{50}$ based on wet weight was 31 mg PFOS/L (22–36 mg PFOS/L) with a NOEC of 6.6 mg PFOS/L

(Boudreau et al. 2003a). At the greatest concentration tested (160 mg PFOS/L), plants exhibited a high percentage of chlorosis as well as frond necrosis.

In an outdoor microcosm study, the toxicity of PFOS to *Myriophyllum sibiricum* and *M. spicatum* was evaluated (Hanson et al. 2005). Both species were exposed to PFOS concentrations ranging from 0.03 to 30 mg PFOS/L for 42 d, and endpoints included plant length, root number and length, node number, and biomass as measured by wet and/or dry weight (see Table 6). *For M. spicatum*, toxicity was observed at PFOS concentrations greater than 3 mg PFOS/L, with EC_{50} being greater than 12 mg PFOS/L. The NOEC was consistently ≥11.4 mg PFOS/L. For *M. sibiricum*, toxicity was observed at PFOS concentrations greater than 0.1 mg/L, and EC_{50} was greater than 1.6 mg PFOS/L. The NOEC was 0.3 mg PFOS/L. Based on these results, *M. sibiricum* was determined to be more sensitive than *M. spicatum*.

Invertebrates. Several acute PFOS toxicity studies have been conducted with the cladoceran *Daphnia magna*, a representative of aquatic invertebrates that is commonly used in standardized toxicity testing. In these acute studies, cladocerans were exposed to various concentrations of PFOS for 48 hr, and mortality and immobility were used as endpoints to calculate LC_{50} and EC_{50} values (Table 7). In one study with *D. magna*, the 48-hr LC_{50} based on mortality and/or immobility was 59 mg PFOS/L (Drottar and Krueger 2000b). Additional acute toxicity tests have been performed with several *Daphnia* species following ASTM guidelines (Boudreau et al. 2003a). In those studies, the 48-hr LC_{50} for *D. magna* was 130 mg PFOS/L and the 48-hr LC_{50} for *D. pulicaria* was 169 mg PFOS/L (Table 7). Based on immobility of the cladocerans, the 48-hr EC_{50} values for *D. magna* and *D. pulicaria* were determined to be 67.2 and 134 mg PFOS/L, respectively (Table 7). Overall, *D. magna* appeared to be more sensitive than *D. pulicaria* based on 48-hr immobility measures (Boudreau et al. 2003a).

In another acute toxicity study with the freshwater mussel *Unio complamatus*, mussels were exposed to various concentrations of PFOS for 96 hr. The 96-hr LC_{50} was determined to be 57 mg PFOS/L while the 96-hr no observed adverse effect concentration (NOAEC) based on mortality was 20 mg PFOS/L (Drottar and Krueger 2000c) (see Table 7). Mussel tissues were analyzed for PFOS content. Chemical analysis of tissue showed that no mortality was associated with PFOS concentrations ≤7.3 mg/kg ww after 96-hr of exposure. In contrast, after 96 hr of PFOS exposure, 90% mortality was observed in mussels containing ≥88.8 mg PFOS/kg.

Toxicity of PFOS to the midge *Chironomus tentans* was evaluated in a 10-d acute test and a 20-d chronic test (see Table 7) (MacDonald et al. 2004). In the 10-d acute study, exposure to PFOS concentrations ranging from 1 to 150 μg PFOS/L resulted in a 30% reduction in survival in the 150-μg PFOS/L treatment whereas no significant mortality was observed

Table 7. Acute and Subacute Toxicity of PFOS to Aquatic Invertebrates.

Species	Test duration	Endpoint	NOAEC (mg/L)[a]	EC_{50}/LC_{50} (mg/L)	Reference
Daphnia magna	48 hr	Survival	33.1 (32.8–34.1)	130 (112–136)	Boudreau et al. 2003a
	48 hr	Immobility	0.8 (0.6–1.3)	67.2 (31.3–88.5)	Boudreau et al. 2003a
	48 hr	Survival/immobility	32	59 (32–87)	Drottar and Krueger 2000b
	48 hr	Survival	NR	58 (46–72)	Robertson 1986
				67 (48–92)	
	21 d	Adult survival	12	15.8[b] (12–24)	Drottar and Krueger 2000f
	48 hr	2nd generation survival	12	NR	
Daphnia pulicaria	21 d	Adult survival	5.3 (2.5–9.2)	42.9 (31.7–56.4)	Boudreau et al. 2003a
	48 hr	Survival	46.9 (33.1–65.3)	169 (136–213)	Boudreau et al. 2003a
	48 hr	Immobility	13.6 (2.2–33.1)	134 (103–175)	
Crassostrea virginica	96-hr	Shell growth	1.8	>2.9	Drottar and Krueger 2000e
Unio complamatus	96 hr	Survival	20	57 (49–65)	Drottar and Krueger 2000c
Artemia salina	48 hr	Survival	NR	9.4 (7.4–12.1)	Robertson 1986
				9.4 (7.3–12.2)	
				8.9 (6.7–11.9)	
Mysidopsis bahia	96 hr	Survival	1.1	3.5 (2.9–4.4)	Drottar and Krueger 2000d
	35 d	Growth, no. young produced	0.24	NR	Drottar and Krueger 2000g
Chironomus tentans	96 hr	2nd generation survival	0.53	NR	MacDonald et al. 2004
	10 d	Survival	0.05	>0.15	
	10 d	Growth	0.05	0.087	
Chironomus tentans	20 d	Survival	0.0	0.092	
	20 d	Growth	0.0	0.094	
Zooplankton	35 d	Community structure	3.0	NR	Boudreau et al. 2003b

[a] 95% confidence intervals in parentheses.
[b] LC_{50} value calculated by the binomial distribution method. NR, not reported.

at lesser treatments. The EC_{50} based on growth was 87 µg PFOS/L and the NOEC was 49.1 µg PFOS/L.

In the 20-d chronic test with midges exposed to nominal concentrations of 1, 5, 10, 50, and 100 µg PFOS/L, a statistically significant reduction in survival was observed at 50 µg PFOS/L whereas no midges survived at 100 µg PFOS/L. The EC_{50} based on a reduction in growth was 94 µg PFOS/L with an EC_{10} of 88 µg PFOS/L. Time to first emergence, rate of emergence, and total emergence were reduced from control levels for all PFOS treatments. The EC_{50} for total emergence was 95 µg PFOS/L and the EC_{10} was 89 µg PFOS/L. A concentration-dependent decline was also observed in the total number of egg masses produced by females when compared to controls. The number of egg masses ranged from 12 in the control group to 1 egg mass in the group exposed to 50 µg PFOS/L. However, no consistent PFOS-related effects on mean number of eggs per egg mass or in percentage hatch were noted in the study.

The results of the 20-d life-cycle study need to be critically evaluated in light of the fact that doses were only measured at the end of the study and measured doses deviated substantially from nominal concentrations. In addition, PFOS has a steep dose–response curve, but this study does not allow appropriate definition of the threshold given the dosing intervals. There are the large differences between NOEC and lowest observed effect concentration (LOEC) values. Moreover, although one would expect the NOEC and EC_{10} values to be similar, the estimated 20-d EC_{10} for growth was 88 mg PFOS/L whereas the NOEC was 22 mg PFOS/L, a 4-fold difference. The difference was even greater for total emergence, where the 20-d EC_{10} was 89.3 mg PFOS/L and the NOEC was <2.3 mg PFOS/L, a 39-fold difference. Given the large dosing interval and the inability to verify the doses given in the end-of-study measurements, data from the life-cycle study most likely do not represent the actual thresholds for adverse effects (Chapman et al. 1996). Therefore, results from the 20-d test will not be used to derive aquatic benchmarks. However, the acute values from this study can be used, given that the 10-d EC_{50} is in agreement with the results from a previous mesocosm study where no midges exposed to >300 µg PFOS/L survived past 10 d (Boudreau et al. 2003b).

The toxicity of PFOS toward marine invertebrates has also been evaluated (see Table 7). In a series of 48-hr toxicity tests with brine shrimp (*Artemia salina*), the average LC_{50} was 9.23 mg PFOS/L (Robertson 1986). In an acute toxicity test with the saltwater mysid *Mysidopsis bahia*, the 96-hr LC_{50} was 3.5 mg PFOS/L and the NOEC was determined to be 1.1 mg PFOS/L based on mortality data (Drottar and Krueger 2000d). The effect of PFOS on a benthic marine invertebrate has also been evaluated. Shell deposition in the eastern oyster *Crassostrea virginica* was examined, and at 1.8 mg PFOS/L, shell growth was inhibited by 20% compared to controls (Drotter and Krueger 2000e). However, an EC_{50} could not be calculated in this study because growth was only inhibited by 28% at the greatest possible PFOS concentration tested (2.9 mg/L) due to limited saltwater solubility (Table 7). Based on these data, it would appear that in short-term studies marine invertebrates are more sensitive to PFOS exposure than freshwater invertebrates.

Life-cycle tests with *D. magna* have been conducted to evaluate the chronic toxicity of PFOS toward aquatic invertebrates. In one study, the 21-d LC_{50} was determined to be 42.9 mg PFOS/L, and the NOEC, based on adult survival, was estimated to be 5.3 mg PFOS/L (Boudreau et al. 2003a). In a second life-cycle toxicity test (Drottar and Krueger 2000f), *D. magna* were exposed to 0, 1.4, 2.8, 5.4, 12, 23, and 46 mg PFOS/L. The 21-d EC_{50} and NOEC were calculated based on adult survival as 12.3 mg PFOS/L and 12 mg PFOS/L, respectively. In a 35-d life-cycle toxicity test with the saltwater mysid, the NOAEC based on growth and number of young produced was 0.24 mg PFOS/L (Drottar and Krueger 2000g) (see Table 7). In the course of the life-cycle tests with both *D. magna* and the saltwater mysid, the young produced were briefly exposed to the same concentrations to which the respective first-generation adults were exposed. Survival was monitored for 48 hr (*D. magna*) or 96 hr (*M. bahia*). After 48 hr of exposure, the results of the daphnid second-generation acute exposure indicated an NOEC of 12 mg PFOS/L. The second generation mysid shrimp were exposed to 0 (negative control) or 0.055, 0.12, 0.24, and 0.53 mg PFOS/L for 96 hr. Survival was ≥95% for all second-generation mysids exposed to these test concentrations. The mysid second-generation acute-exposure NOEC was therefore 0.53 mg PFOS/L.

The effects of PFOS on invertebrate communities have been evaluated in a freshwater microcosm. In this study, 0, 0.3, 3, 10, and 30 mg PFOS/L were administered to a zooplankton community in a controlled freshwater microcosm for 35 d. Results indicated that zooplankton community structure was significantly altered by exposure to 10 and 30 mg PFOS/L. By day 35, the total number of zooplankton species decreased by an average of 45.1% and 74.3% in the 10- and 30-mg PFOS/L treatments, respectively. Therefore, the NOAEC based on changes in zooplankton community structure was determined to be 3.0 mg PFOS/L (Boudreau et al. 2003b) (see Table 7). Within this community, the rank order of zooplankton susceptibility to PFOS was Copepoda > Cladocera > Rotifera (Sanderson et al. 2002).

Fish. Acute toxicity studies with PFOS have been conducted on fathead minnows (*Pimephales promelas*), sheepshead minnows (*Cyprinodon variegatus*), bluegill sunfish (*Lepomis macrochirus*), and freshwater and marine rainbow trout (*Oncorhynchus mykiss*). The results from these acute studies are presented in Table 8. Of the freshwater exposures, the fathead minnow was the most sensitive species with a 96-hr LC_{50} of 9.1 mg PFOS/L and an NOEC of 3.2 mg PFOS/L. After 96 hr of exposure, the sublethal effect of erratic swimming was noted in fathead minnows exposed to concentrations

Table 8. Acute and Chronic Toxicity of PFOS to Fish and Amphibians.

Species	Protocol	NOEC (mg/L)	LOEC (mg/L)	LC_{50}[a] (mg/L)	Reference
Rainbow trout	96-hr static renewal in freshwater	NR	NR	7.8 (6.2–9.8) 9.9 (7.5–13.4)	Robertson 1986
Rainbow trout	96-hr static renewal in freshwater	6.3	13.0	22 (18–27)	Palmer et al. 2002a
Fathead minnow	96-hr static renewal in freshwater	3.2	5.4	9.1 (7.7–11)	Drottar and Krueger 2000h
Fathead minnow	28-d microcosm	0.3	3.0	7.2 (5.2–9.2)	Oakes et al. 2005
Rainbow trout	96-hr static renewal in saltwater	NR	NR	13.7 (10.7–17.7) 13.7 (10.7–17.8)	Robertson 1986
Sheepshead minnow	96-hr static renewal in saltwater	<15	NR	>15	Palmer et al. 2002b
Fathead minnow	Early life stage, 47-d flow-through in freshwater	0.29	0.58	NR	Drottar and Krueger 2000i
Leopard frog	16-wk partial life cycle	0.3	3	6.2 (5.12–7.52)	Ankley et al. 2004
Xenopus	96-hr FETAX[b]	4.82	7.97	15.6	Palmer and Krueger 2001

NR, not recorded as an endpoint in the study.
[a]95% confidence intervals in parentheses.
[b]NOEC and LOEC values taken from the most conservative test results based on growth.

equal to or greater than 5.6 mg PFOS/L (Drottar and Krueger 2000h). Two acute toxicity tests with PFOS have been performed with rainbow trout in freshwater (Robertson 1986; Palmer et al. 2002a). Although the LC_{50} value for PFOS in rainbow trout differed more than twofold between these two studies (see Table 8), the LC_{50} of 22 mg PFOS/L as reported in the Palmer et al. (2002a) study is more reliable than the LC_{50} value reported by Robertson (1986). This conclusion relies on the fact that the LC_{50} value in the Palmer study (2002a) was based on measured PFOS concentrations whereas the Robertson study (1986) was based on nominal concentrations.

The sheepshead minnow inhabits brackish water or saltwater. The acute exposure study with sheepshead minnows was conducted at only one concentration of PFOS, 15 mg PFOS/L (see Table 8); this was the greatest concentration attainable in saltwater and required the addition of methanol (0.05%). No mortality was observed at this concentration after 96 hr, therefore, the 96-hr LC_{50} was reported as >15 mg PFOS/L and the NOEC for sublethal effects as <15 mg PFOS/L (Palmer et al. 2002b). In another study, freshwater rainbow trout were acclimated over 5 d to a final salinity of 30 parts per thousand (‰) and exposed to PFOS for 96 hr (Robertson 1986). For rainbow trout exposed to PFOS in saltwater, the 96-hr LC_{50} was calculated as 13.7 mg PFOS/L, and no sublethal effects were observed among rainbow trout at any PFOS concentration in this study. It should be noted that PFOS concentrations were not measured in this study and that some of the nominal exposure concentrations were greater than the solubility of PFOS in saltwater.

Chronic data are available for the fathead minnow (*Pimephales promelas*) (Drottar and Krueger 2000i). An early life stage toxicity test was conducted with fathead minnows in which eggs and larvae were exposed to PFOS in a flow-through system for 47 d (see Table 8). Measured water concentrations of PFOS in the various treatments were Limit of Quantitation <LOQ, 0.15, 0.30, 0.60, 1.2, 2.4, and 4.6 mg PFOS/L. Minnows exposed to PFOS at concentrations ≤0.30 mg PFOS/L showed no significant reduction in time to hatch, hatching success, survival, or growth. PFOS did not affect percent hatch or growth of fry at any of the concentrations tested. Survival was the most sensitive endpoint in this study. Compared to controls, percent survival was significantly reduced among fathead minnows exposed to concentrations greater than or equal to 0.60 mg PFOS/L. Thus, in this study, the NOAC and LOAC for fathead minnows were determined to be 0.30 and 0.60 mg PFOS/L, respectively (Drottar and Krueger 2000i).

Amphibians.   The Frog Embryo Teratogenesis Assay–Xenopus (FETAX) has been conducted to determine the possible developmental effects of PFOS on African-clawed frogs (*Xenopus laevis*) (Palmer and Krueger 2001). In this assay, survival, growth, and developmental anomalies in frog embryos and tadpoles were examined when exposed to 0 (control), 1.82, 3.07, 5.19, 8.64, 14.4, and 24.0 mg PFOS/L for 96 hr during the early stages of development. Significant early life stage mortality occurred at concentrations ≥14.4 mg PFOS/L (see Table 8). Over three replicate assays, the range of reported LC_{50} values was 13.8–17.6 mg PFOS/L. There was a correlation between PFOS exposure and malformations in each of the three assays. The most commonly observed malformations were improper gut coiling, edema, and notochord and facial abnormalities. The 96-hr EC_{50} values based on malformations ranged from 12.1 to 17.6 mg PFOS/L across all three assays. Tadpole growth was affected in the second and third assays, and the minimum concentrations inhibiting growth were determined to be

7.97 and 8.64 mg PFOS/L. Based on growth, the NOAEC was determined to be 5.19 mg PFOS/L. The teratogenic index (TI), defined as the 96-hr LC_{50} divided by the 96-hr EC_{50}, provides an estimate of the teratogenic risk associated with PFOS. The TIs for the three studies ranged from 0.9 to 1.1 (Palmer and Krueger 2001). A TI in this range would indicate that exposure to PFOS presents a low risk for developmental effects in early-stage African clawed frogs.

The survival and development, from early embryogenesis through complete metamorphosis was evaluated in a study with northern leopard frogs (*Rana pipiens*) exposed to PFOS via water (Ankley et al. 2004). In tadpoles exposed to 0.03, 0.1, 0.3, 1.0, 3.0, and 10 mg PFOS/L, mortality was observed within 2 wk of study initiation in the 10-mg PFOS/L treatment, with greater than 90% mortality being observed by week 4. Tadpole survival was not affected in any other treatment group (see Table 8). The mean LC_{50} at week 5 was 6.21 mg PFOS/L (5.12–7.52 mg PFOS/L). No statistically significant effects were noted for tadpoles from PFOS treatments ≤1.0 mg PFOS/L. However, there was a slight increase in time to metamorphosis and a decrease in total length of tadpoles from the 3.0-mg PFOS/L treatment. In addition, there was a slight increase in the incidence of thyroid follicle cell atrophy that was subtle and difficult to quantitate. The PFOS-related effects in leopard frogs occurred within a concentration range that has been shown to cause effects in fish and invertebrates.

B. Terrestrial

Plants. PFOS toxicity to seven species of crop plants has been determined (Brignole et al. 2003). Onion, ryegrass, alfalfa, flax, tomato, soybean, and lettuce seeds were planted in artificial loam soil mixed with different concentrations of PFOS. Germination studies were conducted for 21 d, after which emergence, survival, and shoot height and weight were determined. The EC_{50} value for survival for all plant species ranged from 57 to >1,000 mg PFOS/kg (Table 9). The 21-d NOAEC based on emergence ranged from 62.5 to 1,000 mg PFOS/kg for all seven species of plants. The NOAEC based on height measurements and shoot weight ranged from <3.91 to 62.5 mg PFOS/kg among all the tested plant species. In general, based on EC_{50} and NOAEC values for survival and emergence, the onion was the most sensitive plant species tested whereas the soybean was the least sensitive. Based on survival data, the order of sensitivity, from most sensitive to least, among these plants was onion > tomato > flax > ryegrass > lettuce > alfalfa > soybean.

Invertebrates. Acute oral and contact toxicity evaluations of PFOS were conducted using the honeybee (*Apis mellifera*) (Wilkins 2001a,b). Honeybees are used in toxicity testing because they are important pollinators of various agricultural crops and are a common model species used to

Table 9. Toxicity of PFOS to Terrestrial Plants.[a,b]

Species	21-d emergence NOAEC (mg/kg)	21-d emergence EC$_{50}$ (mg/kg)	21-d survival NOAEC (mg/kg)	21-d survival EC$_{50}$ (mg/kg)	21-d shoot height NOAEC (mg/kg)	21-d shoot height EC$_{50}$ (mg/kg)
Onion	62.5	210	15.6	57	15.6	47
Ryegrass	62.5	340	62.5	310	3.91	130
Alfalfa	250	745	62.5	450	62.5	250
Flax	250	600	62.5	230	62.5	140
Lettuce	250	560	62.5	390	<3.91	40
Soybean	1,000	>1,000	1,000	>1,000	62.5	460
Tomato	250	470	15.6	105	15.6	94

[a]All data taken from Brignole et al. (2003).
[b]All soil PFOS concentrations are given as means on a wet weight basis (average soil moisture content was 15%).

Table 10. Acute and Chronic Toxicity of PFOS to Terrestrial Invertebrates.

Species	Protocol	NOAEC	LOAEC	LD$_{50}$[a]	Reference
Honeybee (Apis mellifera)	72-hr oral	0.21 µg/bee	0.45 µg/bee	0.40 µg/bee (0.33–0.48)	Wilkins 2001a
Honeybee (Apis mellifera)	96-hr contact	1.93 µg/bee	4.24 µg/bee	4.78 µg/bee (3.8–5.8)	Wilkins 2001b
Earthworm[b] (Eisenia fetida)	14-d soil substrate	77 mg/kg	141 mg/kg	373 mg/kg (316–440)	Sindermann et al. 2002

[a]95% confidence intervals in parentheses.
[b]Soil PFOS concentrations given on a dry weight basis (average soil moisture content was 32%).

evaluate the toxicity of pesticides. In an oral toxicity study, worker honeybees were administered 0.21, 0.45, 0.99, 2.17, and 4.78 µg PFOS/bee in an aqueous sucrose solution for 4 hr. No mortality or sublethal effects were noted in the 0.21-µg PFOS/bee treatment. A steep dose response was noted between mean intakes of 0.45 and 2.2 µg PFOS/bee. The 72-hr oral LD$_{50}$ was calculated to be 0.40 µg PFOS/bee with a slope of 4.8 (Table 10). If the LD$_{50}$ is converted to dose per kilogram of food, the value is 2.0 mg PFOS/kg sugar solution. Based on the 72 hr oral LD$_{50}$, PFOS was classified as "highly toxic" by the International Commission for Bee Botany (ICBB) (Wilkins 2001a).

In the acute contact study, treatments were administered by applying 1.93, 4.24, 9.3, 20.5, and 45.0 µg PFOS/bee to the thorax of bees. Based on mortality, the NOEL was 1.93 µg PFOS/bee while a steep dose response was observed between the 4.24- and 9.30-µg PFOS/bee doses. The slope of the dose–response line was estimated to be 4.0 and the 96-hr LD_{50} was determined to be 4.8 µg PFOS/bee (see Table 10). No dose-related sublethal effects were noted at any dose level in the study. Based on the 96-hr contact LD_{50}, PFOS was classified as moderately toxic by the ICBB (Wilkins 2001b).

Studies of the acute toxicity of PFOS to the earthworm *Eisenia fetida* have been conducted (Sindermann et al. 2002). Earthworms are closely associated with the soil environment and can therefore be good bioindicators of contaminant loads in soils. PFOS was incorporated into an artificial soil substrate at five different concentrations, and adult earthworms were exposed to the soil for 14 d. On day 14, surviving worms from each treatment group were placed on moistened filter paper and allowed to purge their gut contents for 24 hr. The entire mass of worms from each treatment group was then prepared for analysis of PFOS concentrations. Throughout the study, worms were evaluated for mortality, body weight, and burrowing behavior. The 14-d LC_{50} was determined to be 373 mg PFOS/kg soil (see Table 10). At the two greatest PFOS concentrations tested, 488 and 1042 mg/kg in soil, the worms exhibited significant mortality, aversion to burrowing into the soil, and significant losses of body weight. The 14-d NOAEL, based on survival, was determined to be 77 mg PFOS/kg soil. PFOS was present in the tissues of the earthworm following 14 d of exposure. No mortality was observed, and there was no effect on either burrowing behavior or weight at tissue concentrations ≤195 mg PFOS/kg (wet tissue weight).

C. Terrestrial Vertebrates

Avian Acute Studies. Acute feeding studies have been conducted using 10-d-old juvenile mallards (*Anas platyrhynchos*) and northern bobwhite quail (*Colinus virginianus*) (Gallagher et al. 2004a,b). The nominal dietary concentrations of PFOS were 0 (control), 8.8, 17.6, 35.1, 70.3, 141, 281, 562, and 1,125 mg/kg feed. After exposure to contaminated feed for 5 d, birds were offered untreated feed for another 17 d and the studies were terminated on day 22. For mallards, exposure to PFOS at dietary concentrations ≤141 mg PFOS/kg did not result in mortality or overt signs of toxicity. The dietary 8-d LC_{50} was determined to be 603 mg PFOS/kg (Table 11). Based on average daily intake of PFOS, the 8-d LD_{50} for mallards was determined to be 150 mg PFOS/kg bw/d. Based on the reduction of body weight at day 8, the NOAEL was determined to be 35.1 mg PFOS/kg feed and the LOAEL was 70.3 mg PFOS/kg feed. PFOS was present in the liver and serum of mallard ducks after the exposure period. The half-life of PFOS in mallard blood serum and liver was estimated to be 6.86 and 17.5 d, respectively.

For juvenile quail, no overt signs of toxicity or mortality were observed at concentrations ≤70.3 mg PFOS/kg feed. The dietary 8-d LC_{50} was determined to be 212 mg PFOS/kg, and the 8-d LD_{50} based on average daily

Table 11. Acute Toxicity of PFOS to Avian Species.

Species	Protocol	Matrix	NOAEC (mg/kg)	LOAEC (mg/kg)	LC$_{50}$[a] (mg/kg)	Reference
Mallard	8-d	Diet	35.1	70.3	603 (430–919)	Gallagher
(*Anas*	dietary	Serum	30.5	48.1	–	et al.
platyrhynchos)		Liver	15.0	29.7	–	2004a
Northern	8-d	Diet	70.3	141	212 (158–278)	Gallagher
bobwhite quail	dietary	Serum	41.2	41.5	–	et al.
(*Colinus*		Liver	44.0	70.3	–	2004b
virginianus)						

[a]95% confidence intervals in parentheses.

intake was 61 mg PFOS/kg bw/d. Based on treatment-related mortality, signs of toxicity, and body weight gain, the NOAEL was determined to be 70 mg PFOS/kg feed and the LOAEL to be 141 mg PFOS/kg feed (see Table 11). PFOS was present in the liver and serum of bobwhite quail after the exposure period. The half-life of PFOS in liver was estimated to be 12.8 d, whereas the half-life of PFOS in blood serum could not be estimated.

Avian Chronic Toxicity Studies. Reproduction studies were conducted with adult mallard ducks and bobwhite quail that were fed at 0, 10, 50, or 150 mg PFOS/kg feed for up to 21 wk (Gallagher et al. 2003a,b). Adult health, body and liver weight, feed consumption, gross morphology and histology of body organs, and reproduction were examined. However, several weeks into the study birds from the 50- and 150-mg PFOS/kg dietary treatments experienced a host of adverse effects, including high rates of mortality, and these treatment groups were euthanized by week 7 of the studies (Table 12). Exposure of adult mallards or quail to 10 mg PFOS/kg in the diet did not result in overt signs of toxicity or treatment-related effects on body weight, body weight gain, and feed consumption. In addition, no treatment-related effects were noted on liver weight of adult female or male mallards from the 10-mg PFOS/kg dietary treatment. Although no treatment-related effects on liver weight of male quail were noted, there was a slight but statistically significant PFOS-related effect on liver weight in females. However, when liver weights were normalized to body weight, no differences were observed between treated and control female quail. In mallards, no statistically significant differences were noted on any reproductive parameters in the 10-mg PFOS/kg dietary treatment when compared to controls. For quail, no treatment-related effects were noted on egg production or embryo viability in the 10-mg PFOS/kg dietary treatment group whereas there were slight but not statistically significant reductions in egg fertility and hatchability in the 10-mg PFOS/kg diet group. This reduction was accompanied by a significant reduction in the number of 14-d-old survivors as a percentage of the number of eggs set (see Table 12).

Table 12. Measurement Endpoints and Associated Dietary NOAEL and LOAEL Values for PFOS in a Chronic Study with Mallards, Northern Bobwhite Quails, and Their Offspring.[a,b]

	Mallard		Northern bobwhite	
Endpoint	NOAEL (ppm)	LOAEL (ppm)	NOAEL (ppm)	LOAEL (ppm)
Adult				
Mortality	10	50	10	50
Body weight	10	50	10	50
Feed consumption	50	150	10	50
Liver weight				
Males	10	–	10	–
Females	10	–	10	–
Gross pathology				
Males	<10	10	10	–
Females	10	–	10	–
Histopathology	10	–	10	
Reproductive	10	–		10
Offspring				
14-d survivability	10	–		10[c]
Hatchling/juvenile body weight	10	–	10	–
Juvenile liver weight	10	–	10	–

[a]All concentrations reported on a wet weight basis.
[b]10 ppm group exposed for 21 wk; 50 ppm group exposed for 7 wk. Adult liver weight, gross pathology, histopathology, and offspring endpoints were evaluated only in the control and 10-mg PFOS/kg treatments at study termination.
[c]LOAEL based on increased offspring survival.

For both female mallards and quail treated with 10 mg PFOS/kg feed, no treatment-related effects were noted on gross pathological or histopathological endpoints. However, in adult male mallards and quail treated in 10 mg PFOS/kg feed, there was a greater incidence of reduced testis size. No effects on spermatogenesis were observed in male quail with small testes, but some morphological alterations were observed in mallards. Specifically, several of the mallards with small testes also exhibited a reduction in spermatogenesis that was characterized by fewer maturing/mature spermatozoa in the tubules. Taken together with the fact that other treated mallards with small testes did not exhibit any alteration in spermatogenesis, PFOS may have accelerated early postreproductive phase regression in exposed birds, a normal physiological phenomenon. In many seasonal breeders, males display a cyclic reduction in fertility during nonbreeding season. This reduction in testis size is accompanied by regression of testicular germinal epithelium, leading to a decline in spermatozoa production and testicular size (Rosentrauch et al. 1994; Wikelski et al. 2003). Because no effects were noted on egg production or fertilization, the toxicological significance of small testes in males did not appear to significantly affect reproductive performance in the males exposed to PFOS in the diet.

PFOS was accumulated into blood serum and liver in birds exposed to PFOS in the diet. At study termination, postreproductive liver and serum samples from adult male birds were greater than those measured in adult females taken from the same treatment groups. Despite the sex-related differences in serum and liver PFOS concentrations, the ratios of PFOS concentrations in serum to those in the liver were similar for both sexes, with an average ratio of 1.7 and 1.5 for quail and mallards, respectively. In a time-course study, adult male and female quail accumulated PFOS into serum in a similar manner and magnitude until the onset of egg laying. However, once egg laying commenced, serum PFOS concentrations in females were reduced to approximately 85% from the prereproductive levels while male serum PFOS levels were unaffected. This phenomenon is of particular importance when evaluating adult female serum PFOS concentrations in that reproductive condition can have a significant influence on measured PFOS concentrations and could lead to erroneous conclusions relative to evaluating the risk posed by PFOS to avian population. In studies to date, no influence of sex has been noted on liver or serum PFOS concentrations in juvenile or immature birds.

Based on reproduction studies, the LOAEL for male and female mallards was 50 mg PFOS/kg in the diet (6.4 mg PFOS/kg bw/d) while the NOAEL was 10 mg PFOS/kg in the diet (1.5 mg PFOS/kg bw/d) (Table 13). For adult female quail, the LOAEL was 10 mg PFOS/kg ww diet (0.77 mg PFOS/kg bw/d) based on a decreased survivorship of 14-d-old quail offspring, although a NOAEL could not be determined. For adult male quail, the LOAEL based on effects on body weight and mortality was 50 mg PFOS/kg diet (2.64 mg PFOS/kg bw/d) whereas the NOAEL was determined to be 10 mg PFOS/kg in the diet (0.77 mg PFOS/kg bw/d).

IV. Hazard Assessment
A. Overview

In many assessment programs, ecological risks posed by contaminants are evaluated in a two-tiered process (USEPA 1997). In the first tier, a screening-level assessment is performed wherein contaminant concentrations in the environment are compared to toxicological benchmarks that represent nonhazardous concentrations to biota. These benchmarks are usually derived using conservative assumptions and are used to screen for chemicals that will be included in subsequent analyses of risk. The second tier, or baseline ecological risk assessment, is a more refined assessment that contains additional information including bioavailability data, site- and species-specific information on exposure, and potential effects for chemicals

Table 13. NOAEL and LOAEL Values in Various Matrices in Adult Mallards and Northern Bobwhite Quails and Their Offspring in a Chronic Dietary Study with PFOS.

	Mallard		Quail	
Measures of PFOS exposure[a]	NOAEL	LOAEL	NOAEL	LOAEL
Adult males				
Dose (ppm)	10	50	10	50
ADI (mg PFOS/kg body weight/d) over 21-wk period	1.49	–	0.77	2.64
Serum (µg PFOS/mL) at study termination (21 wk)	87.3	–	141	–
Liver (µg PFOS/g) at study termination (21 wk)	60.9	–	88.5	–
Adult females				
Dose (ppm)	10	50	–	10[b]
ADI (mg PFOS/kg body weight/d) over 21-wk period	1.49	6.36	–	0.77
Serum (µg PFOS/mL), prereproductive phase (5 wk)	76.9	–	–	84
Serum (µg PFOS/mL), reproductive phase, at study termination (21 wk)	16.6	–	–	8.7
Liver (µg PFOS/g) at study termination (21 wk)	10.8	–	–	4.9
Offspring				
Yolk (µg PFOS/mL)	52.7	–	–	62
Liver (µg PFOS/g)[c]	3.39	–	–	5.5
Serum (µg PFOS/mL)[c]	4.69	–	–	12.5

[a]Effect values for diet and average daily intake (ADI) are reported as dietary concentrations. Serum and liver effect values are reported as measured tissue values. For convenience, all effect values in the table are reported as measured values (or level). All concentrations reported on a wet weight basis. Only the control and 10 ppm treatments were maintained through study termination (week 21).

[b]LOAEL for adult females was based on a decrease in the 14-d-old survivability of offspring.

[c]Offspring liver and serum LOAEL values are averages of female and male concentrations.

identified in the screening-level assessment. In this type of analysis, toxicological benchmarks are one of several lines of evidence that are used to support or refute the potential of ecological effects. In this chapter, we derived benchmark values that can be used in a Tier II screening assessment. That is, these benchmarks represent screening values that can be used to categorize potential risks to target biota from environmental concentrations of PFOS, but are not intended for use as standards or to quantify these risks as would occur in a Tier I baseline assessment.

The conservative nature of benchmark values is partly because of the presence of data gaps related to the interpretation and extrapolation of toxicity data between species, the use of different laboratory toxicological endpoints, experimental design, and extrapolation of laboratory data to natural systems (Duke and Taggart 2000). For example, most toxicity data used to derive benchmarks are from toxicity tests conducted in the laboratory with a few surrogate species that may not share similar life histories or feeding strategies with the species of concern. To account for these potential differences, uncertainty factors (UF), sometimes referred to as safety factors, are estimated and applied to the data such that a conservative benchmark is derived that will be protective of a population which is potentially exposed to a stressor. As additional data become available, the magnitude of the uncertainty decreases and a more accurate estimate of the actual toxic threshold for a species is achieved. As a result, due to the conservative assumptions used to derive the benchmark values, these values represent environmental concentrations below which adverse effects to target populations would not be expected to occur. Furthermore, although exceedance of the benchmarks does not indicate any particular level or type of risk posed by the chemical to populations, it can be used to identify those situations where additional studies are needed to better assess risks.

Benchmark values typically include the effects of both short-term acute exposures and longer-term chronic exposures for species of concern. Information on the accumulation kinetics, disposition, acute lethality, and recovery are useful to assess the potential effects of short-term exposures at contaminated sites. Information from chronic exposures typically includes ecologically sensitive and relevant endpoints such as survival, growth, and reproduction.

The following discussion presents benchmark values for aquatic organisms, including a benchmark for protection of freshwater organisms, a benchmark for protection of aquatic plants, and a critical body burden for fish. The discussion then presents benchmarks for terrestrial plants, terrestrial invertebrates, and avian species.

Multiple approaches have been used to derive benchmark values using available toxicity test data. There are a number of ways to derive benchmarks. This simplest approach is to use data from toxicity tests in the species of concern and use resulting NOEC and LOEC values as benchmarks (Suter 1996). A similar approach is to use toxicity data from a surrogate species and apply application factors to account for potential differences. An alternative method is a combination of multiple tests such as is used in the derivation of U.S. Ambient Water Quality Criteria for the Protection of Aquatic Life (NAWQC) (Stephan et al. 1985). Because the quantity, quality, and type of data available for each taxonomic group evaluated differ, various approaches were used to establish benchmarks that were based on the best available science as well as being protective of targeted populations.

Table 14. Data Requirements and Data Available for the Determination of a Tier I Final Acute Value (FAV).

GLI data requirements	Species tested	Acute result?	Chronic result?
1. Class Osteichthyes, family Salmonidae	*Oncorhynchus mykiss*	√	
2. Class Osteichthyes, second family	*Pimephales promelas*	√	√
3. Phylum Chordata, third family	*Xenopus laevis*	√	
4. Planktonic crustacean	*Daphnia magna*	√	√
5. Benthic crustacean			
6. Insect			
7. A family in a phylum other than Arthropoda or Chordata	*Unio complamatus*	√	
8. A family in any order of insect or any phylum not already represented	*Chironomus tentans*	√	

GLI, Great Lakes Initiative.

B. Aquatic

Aquatic Organisms Benchmark. Available data are insufficient for calculating a national water quality criterion pursuant to EPA methodology. EPA's Great Lakes Initiative (GLI) guidance provides for calculation of screening values when sufficient data for a water quality criterion are not available. Accordingly, the methodology to derive a screening benchmark for protection of aquatic animals was based on the GLI (USEPA 1995).

The GLI provides specific procedures and methodology for utilizing toxicity data to derive water quality values that are protective of aquatic organisms. The GLI presents a two-tiered methodology (Tier I and Tier II). Tier I procedures are essentially the same as the procedures for deriving national water quality criteria. The Tier II aquatic life methodology is used to derive values when fewer toxicity data exist. As there are greater uncertainties associated with limited toxicity data, the Tier II methodology generally produces more stringent values than the Tier I methodology. EPA has indicated that Tier II values are not intended to be adopted as state water quality standards (USEPA 1995).

A Tier I freshwater value requires the availability of acute data from at least one species in each of eight listed families and chronic data from at least three. As shown in Table 14, the PFOS dataset includes acute data from six of the eight required families and chronic data from two. Thus, a Tier I value cannot be calculated. However, a Tier II value can be derived. A Tier II screening value protective against chronic effects can be calculated using the available acute data and an acute-to-chronic ratio.

Secondary Chronic Value (SCV). To calculate the Tier II freshwater Secondary Chronic Value (SCV), the guidance requires use of the least acute toxicity value. For PFOS, the least value for a freshwater organism was an EC_{50} for growth of 0.087 mg PFOS/L from the 10-d definitive study in *Chironomus* sp. This value is lower by a factor of approximately 40 than the next most sensitive aquatic organisms.

This least acute toxicity value is then divided by a Secondary Acute Factor (SAF) to estimate a Secondary Acute Value (SAV) for PFOS. The SAF is an adjustment factor specified by the guidance. In this case, where data are available for six of the eight required families, the prescribed SAF is 5.2.

Thus, the SAV is calculated as follows (Eq. 1):

$$\text{secondary acute value (SAV)} = \frac{\text{lowest acute value}}{\text{secondary acute factor (SAF)}}$$

$$= \frac{0.087\,\text{mg/L}}{5.2} = 0.017\,\text{mg PFOS/L} \qquad (1)$$

This secondary acute value is then adjusted to a chronic value using an acute-to-chronic ratio. Acceptable acute and chronic data are available for *Daphnia magna* (*Daphnia* sp.), fathead minnows (*P. promelas*), and mysid shrimp (*Mysis* sp.). Although data from saltwater organisms are not used to determine the freshwater acute benchmark (SAV), under the GLI guidance, data from saltwater organisms such as mysid shrimp may be used to estimate an acute-to-chronic ratio (ACR). The Final Acute to Chronic Ratio (FACR) is calculated as the geometric mean of the three acceptable ACRs. The FACR for PFOS was determined to be 13.9.

The Secondary Chronic Value (SCV) was calculated dividing the SAV by the FACR. The SCV for PFOS was calculated as follows (Eq. 2):

$$\text{secondary chronic value (SCV)} = \frac{\text{secondary acute value (SAV)}}{\text{final acute to chronic ratio (FACR)}}$$

$$= \frac{0.017\,\text{mg/L}}{13.9} = 0.0012\,\text{mg PFOS/L} \qquad (2)$$

Thus, the Tier II screening benchmark for aquatic organisms was determined to be 1.2 µg PFOS/L. This value is conservative given that the EC_{50} value for *Chironomus* is lower by a factor of about 40 than the next most sensitive species.

Aquatic Plant Benchmark

Screening Plant Value. The Screening Plant Value (SPV) represents the least concentration from a toxicity test with an important aquatic plant species where the concentrations of test material have been measured and the endpoint monitored in the study is biologically important. For PFOS,

Table 15. Selected Freshwater Acute and Chronic Plant Data for the Derivation of a Screening Plant Value (SPV).[a]

Species	Acute value (EC$_{50}$)	SMAV[b] (mg/L)	NOEC-LOEC (mg/L)	Chronic value[d] (mg/L)	SMCV (mg/L)
Green alga, Selenastrum capricornutum	71		44–86 (4-d growth)	61.5	
Green alga, Selenastrum capricornutum	48.2	58.5	5.3[c] (4-d biomass)	5.3	18
Green alga, Chlorella vulgaris	81.6	81.6	8.2[c] (4-d biomass)	8.2	8.2
Blue-green alga, Anabaena flos-aquae	131	131	93.8–143 (4-d growth)	115	115
Diatom Navicula pelliculosa	263	263	111–150 (4-d growth)	129	129
Duckweed, Lemna gibba	108		15.1–31.9 (7-d frond number)	21.9	
Duckweed, Lemna gibba	31.1	58	6.6[c] (7-d biomass)	6.6	12
Water milfoil, Myriophyllum spicatum	12.5		2.9–11.4 (42-d biomass)	5.74	
Water milfoil, Myriophyllum sibiricum	2.4	5.5	0.3–2.9 (42-d root length)	0.93	2.3

[a]Toxicity values selected as the most sensitive endpoint for each plant species.
[b]SMAV, species mean acute values SMCV, is the species mean chronic value. Both are calculate as the geometric mean of species specific toxicity values.
[c]NOEC values were calculated by regression techniques such as the IC$_{10}$.
[d]Chronic value is the geometric mean for the NOEC and LOEC. If only a NOEC was reporte for a study, then the chronic value was given as the NOEC.

the freshwater SPV was based on the species mean chronic value (SMCV) for *Myriophyllum* sp. (Table 15). The selection of this study was based on the fact that PFOS water concentrations were measured values and that biological and ecologically important endpoints related to growth were measured in the studies. Using the SMCV for *Myriophyllum* sp. is consistent with the several EPA guidelines, including the GLI. As a result, the screening benchmark for plants was determined to be 2.3 mg PFOS/L.

Critical Body Burden: Fish. The critical body residue (CBR) hypothesis provides a framework for analyzing aquatic toxicity data in terms of its mode of action and tissue residue concentrations (McCarty and Mackay 1993; Di Toro et al. 2000). The key assumption of the hypothesis is that adverse effects are elicited when the molar concentration of a chemical in an organism's tissues exceeds a critical threshold. Typically, in the absence

Table 16. Cumulative Mortality and Whole-Body PFOS Concentrations in Bluegill Exposed to 0.86 mg/L in a Bioconcentration Study.[a]

Exposure day	PFOS (µg/kg)	Number exposed	Cumulative mortality
0.2	2,880	55	0
1.0	9,643	55	0
3.0	33,700	55	0
7.0	81,750	55	0
14	159,000	55	16
21	178,250	55	35
28	241,750	55	52

[a]PFOS concentrations are means for fish sampled on the indicated dates. Concentrations are expressed on a wet weight (ww) basis.

of direct tissue measurements, under steady-state conditions the CBR can be expressed mathematically as the product of the effect concentration in water determined in an aquatic test and the bioconcentration factor. Implicit in this hypothesis is the assumption that a chemical is accumulated in tissues via a partitioning process and it has reached a steady state within the test period. Thus, the CBR is a time-independent measure of effect for organisms exposed to the chemical. One problem with this assumption is that in many cases organisms may not have achieved a steady-state concentration such that using the BCF would overestimate the actual whole-body concentration one would expect during a standard aquatic acute toxicity test. In addition, this model does not take into account accumulation of chemicals into target tissues that may accumulate a chemical in a manner that differs from that observed on a whole-body basis (Barron et al. 2002). Thus, these factors may result in an overestimate of the CBR that would underestimate the risk an aquatic organism would be exposed to in a natural setting.

For this analysis, the problems inherent in deriving a CBR using BCF values can be avoided because actual whole-body concentrations of PFOS in bluegill associated with toxicity are available. To estimate a CBR level for PFOS in fish, we used data from a bluegill bioconcentration study where significant mortality occurred at the higher dose (Drottar et al. 2001). In this study, bluegill sunfish were exposed to 0.086 and 0.87 mg PFOS/L for up to 62 d followed by a depuration period. However, at 0.87 ppm, mortality was noted by day 12 with 100% mortality being observed by day 35. Thus, at this dose, no fish survived to the end of the uptake phase of the study. Mortality and whole-body PFOS concentrations collected during the study at the 0.87-ppm exposure are given (Table 16).

Probit analysis was used to estimate a critical body residue using tissue PFOS concentration as our independent variable and mortality as the dependent variable. Use of probit analysis allowed the calculation of point estimates along the dose–response relationship such that a chronic no-effect threshold based on survival could be estimated from the mortality data (Mayer et al. 1986; McCarty and MacKay 1993). The 28-d LD_{50} based on whole-body concentrations was 172 mg PFOS/kg ww. The 95% lower and upper confidence limits for the LD_{50} were 163 and 179 mg PFOS/kg, respectively. As an estimate of a NOAEL for PFOS-induced mortality in bluegill, we extrapolated down to the LD_{01}, which was 109 mg PFOS/kg ww. The 95% lower and upper confidence limits for the LD_{01} were 87 and 123 mg PFOS/kg ww. Based on the foregoing statistical evaluation of the data, the tissue concentration that would not be expected to cause acute adverse effects in fish is 109 mg PFOS/kg. For reasons of potential differences in species sensitivity, the lower 95% confidence limit of the LD_{01} was used as a conservative estimate of a NOAEL. Based on this analysis, tissue concentrations less than 87 mg PFOS/kg would not be expected to cause acute effects in fish.

C. Terrestrial Toxicity Benchmarks

Benchmark values for terrestrial species were estimated by an approximation method in which the most sensitive species was identified for each taxonomic group and applications factors were used to account for uncertainties (Giesy et al. 2000). For terrestrial plants and invertebrates, chronic benchmarks were estimated from the NOEC of the most sensitive species. If only one species was available to estimate a chronic value, then an uncertainty factor of 2 was applied to the NOEC. If a chronic NOEC was not available, then an uncertainty factor of 3 applied to a chronic LOEC. Finally, when no chronic data were available, a no-mortality level (NML) was derived by applying a fivefold safety factor to the LC_{50}/EC_{50} of the most sensitive species (Urban and Cook 1986).

An approximation method was also used to calculate an avian benchmark value, but a more sophisticated protocol was used to estimate an uncertainty factor that included a weight of evidence approach.

Terrestrial Plants. The plant toxicity study evaluated seven different species and was considered to be a chronic test that evaluated several biologically important endpoints. Based on endpoints related to survival and growth, the onion was determined to be the most sensitive species. However, the effects were observed on the 21-d shoot height of lettuce at <3.91 mg PFOS/kg ww, and as a result this value was used to derive a plant benchmark. Because a chronic NOEC for the most sensitive species and endpoint was not determined in this study, an uncertainty factor of 3 was used to derive a benchmark. Thus, the screening benchmark for terrestrial plants was 1.3 mg PFOS/kg soil ww or 1.5 mg PFOS/kg soil dw.

Terrestrial Invertebrates. For soil organisms, one toxicity test with earthworms is available to estimate a benchmark value. In this 14-d study, no effects were observed on worms exposed to 77 mg PFOS/kg soil dw or less nor were there any effects noted on worms with tissue PFOS concentrations of 195 mg PFOS/kg ww or less. Because a NOEC was available for only one species, an additional uncertainty factor of 2 was used in the analysis. The result is a chronic benchmark of 39 mg PFOS/kg dw soil (or 33 mg PFOS/kg ww soil) and a worm tissue-based benchmark of 98 mg PFOS/kg ww.

Avian Species. Benchmarks for avian species were determined by an approximation approach where the most ecologically important endpoint from the relevant toxicity study is adjusted by an uncertainty factor that takes into account data gaps and extrapolations. Uncertainty factor assignment was conducted using the USEPA Great Lakes Initiative methodology (USEPA 1995). Calculations using European Commission OECD guidance are presented elsewhere (Newsted et al. 2005c) but produce similar results.

In this approach using GLI methodology, three categorical uncertainties are evaluated, including (1) uncertainty with LOAEL to NOAEL extrapolation (UF_L), (2) uncertainty related to duration of exposure (UF_S), and (3) uncertainty related to intertaxon extrapolations (UF_A). Uncertainty factors for each category are then assigned values between 1 and 10 that are based on available scientific findings and best professional judgment (Abt Associates 1995; Chapman et al. 1998).

For the species of interest, the characteristics of a top avian predator such as eagles or cormorants were used in the analysis. These species have been shown to accumulate PFOS to a greater degree than lower trophic level bird species and will integrate potential contributions of PFOS from both aquatic and terrestrial exposure pathways (Ankley et al. 1993; Bowerman et al. 1998; Drouillard et al. 2001). In addition, many of these bird species have been shown to be sensitive to other classes of organic compounds and thus provide an early warning system for the presence and effects of contaminants within aquatic ecosystems (Giesy et al. 1994; Giesy and Kannan 1998).

Based on the data from the quail reproduction study and the characteristics of a level IV avian predator, uncertainty factors (UFs) were assigned to account for data gaps and extrapolations in the analysis (Table 17). TRVs were calculated based on dietary, average daily intake (ADI), egg yolk, and serum and liver PFOS concentrations (Table 18). Dietary, ADI, and egg yolk-based TRVs for the level IV avian predator were 0.28 mg PFOS/kg diet, 0.021 mg PFOS/kg bw/d, and 1.7 μg PFOS/mL yolk, respectively (see Table 18).

Table 17. Assignment of Uncertainty Factors for the Calculation of a Generic Trophic Level IV Avian Predator Toxicity Reference Value (TRV) for PFOS.[a]

Uncertainty factors	Notes
Intertaxon extrapolation (UF$_A$)	The laboratory study used to determine a threshold dose was from northern bobwhite quail. Because this species belongs to the same taxonomic class but different order, UF$_A$ = 6.
Toxicological endpoint (UF$_L$)	The quail study determined a LOAEL but not a NOAEL based on multiple endpoints that included reproduction. Furthermore, the difference between the LOAEL and control was less than 20% for the effected reproductive endpoints. Taken together with other study data, the UF$_L$ = 2.
Exposure duration (UF$_S$)	The quail reproductive study was conducted for 20 wk and evaluated several important life stages including embryonic development and offspring growth and survival, so UFs = 3.
* Overall UF for TRV	UF = 6*2*3* = **36**

[a]Selection of uncertainty factors based on the Great Lake Initiative (USEPA 1995).

Table 18. PFOS Toxicity Reference Value (TRV) Values for a Generic Avian Trophic Level IV Predator Based on Dietary, Liver, Serum, and Egg Yolk Concentrations.[a]

	Male			Female		
Factor	LOAEL	TRV[b]	PNEC[c]	LOAEL	TRV[b]	PNEC[c]
ADI (mg PFOS/kg/d)[d]	0.77	0.022	0.026	0.77	0.022	0.026
Liver (µg PFOS/g, ww)	88	2.5	2.9	4.9	0.14	0.16
Serum (µg PFOS/ml)	141	4.0	4.7	8.7	0.25	0.29
Egg yolk (µg PFOS/ml)				62	1.8	2.1

[a]LOAEL values based on bobwhite quail definitive study.
[b]TRV estimated with total uncertainty factor derived for a generic level 4 predator by the GLI protocol.
[c]Predicted No Effect Concentration (NEC) estimated with total assessment factor of 30.
[d]ADI, average daily intake (mg PFOS/kg, bw/d); estimates were based on pen averages.

Development of TRVs based on liver and serum concentration required consideration of gender differences. Because of sex-specific differences in serum and liver PFOS concentrations at study termination, tissue-based TRVs in males were approximately 17 fold greater than values reported for females. These differences were most likely a result of PFOS being transferred to egg from adult females during egg laying. This conclusion is substantiated by the fact that during the prereproductive phase of the study serum concentrations in females were similar to that observed in males (Newsted et al. 2005b). To derive TRVs that integrate the differences between sexes and reproductive condition, geometric means of male and female serum and liver PFOS concentrations were calculated. Based on geometric means, the TRVs for serum and liver were 1.0 μg PFOS/ml and 0.6 μg PFOS/g ww, respectively. These TRVs thus are protective of all adult birds regardless of their reproductive phase.

Because of the conservative assumptions used the analyses, dietary or tissue concentrations at or less than these TRV values would not be expected to pose significant risks to avian populations. In light of observed no-effect levels in the studies, however, population-level effects would not be expected to occur until a concentration of 6.0 mg PFOS/kg in the diet, 5.0 μg PFOS/g ww in the liver, or 9.0 μg PFOS/ml in the serum was exceeded.

Summary

Based on available toxicity data, protective screening-level concentrations of PFOS were calculated for aquatic and terrestrial organisms. Using the Great Lakes Initiative, water concentrations of PFOS were calculated to protect aquatic plants and animals. The screening plant value (SPV) protective of aquatic algae and macrophytes was calculated as 2.3 mg PFOS/L. The secondary chronic value protective of aquatic organisms was 1.2 μg PFOS/L. The screening-value water concentrations less than or equal to 1.2 μg PFOS/L would not pose a potential risk to aquatic organisms. Because the aquatic benchmark is based on the most sensitive species, this benchmark should also be protective of other aquatic organisms, including amphibians. The tissue-based TRV for fish was determined to be 87 mg PFOS/kg ww.

For terrestrial plants, a screening benchmark was determined to be 1.3 mg PFOS/kg soil ww or 1.5 mg PFOS/kg soil dw, whereas for soil invertebrates such as earthworms the benchmark value was 39 mg PFOS/kg dw soil or 33 mg PFOS/kg ww soil.

For avian species, dietary, ADI, and egg yolk-based benchmarks were determined as 0.28 mg PFOS/kg diet, 0.021 mg PFOS/kg bw/d, and 1.7 μg PFOS/mL yolk, respectively. Benchmarks for serum and liver for the protection of avian species were 1.0 μg PFOS/mL and 0.6 μg PFOS/g ww, respectively. However, no-effect levels in laboratory studies suggest actual population-level effects would not be expected to occur until a concentration of 6.0 mg PFOS/kg in the diet, 5.0 μg PFOS/g ww in the liver, or 9.0 μg PFOS/mL in the serum was exceeded, thus indicating the conservative nature of the benchmarks.

Acknowledgments

The authors thank John Butenhoff for his insightful comments and helpful advice on this manuscript.

References

3M Company (2001a) Screening studies on the aqueous photolytic degradation of potassium perfluorooctane sulfonate (PFOS). Lab Request Number W2775. EPA Docket AR226-1030a041. 3M Environmental Laboratory, St. Paul, MN.

3M Company (2001b) Hydrolysis reactions of perfluorooctane Sulfonate (PFOS). Lab Request Number W1878. EPA Docket AR226-1030a039. 3M Environmental Laboratory, St. Paul, MN.

3M Company (2001c) Solubility of PFOS in natural seawater and an aqueous solution of 3.5% of sodium chloride. Laboratory Project Number E00-1716. EPA Docket AR226-1030a026. 3M Company, 3M Environmental Laboratory, St. Paul, MN.

3M Company (2001d) Solubility of PFOS in octanol. Laboratory Project Number E00-1716. EPA Docket AR226-1030a027. 3M Company, 3M Environmental Laboratory, St Paul, MN.

3M Company (2001e) Solubility of PFOS in water. Laboratory Project Number E00-1716. EPA Docket AR226-1030a025. 3M Company, 3M Environmental Laboratory, St. Paul, MN.

3M Company (2001f) Soil adsorption/desorption study of potassium perfluorooctanesulfonate (PFOS). Laboratory Project Number E00-1311. EPA Docket AR226-1030a030. 3M Company, 3M Environmental Laboratory, St. Paul, MN.

3M Company (2003) Environmental and Health Assessment of Perfluorooctane Sulfonate and its Salts. Available on USEPA Administrative Record AR-226-1486.

Abt Associates (1995) Technical basis for recommended ranges of uncertainty factors used in deriving wildlife criteria for the Great Lakes Water Quality Initiative. Final Report. Office of Water, U.S. Environmental Protection Agency, Washington, DC.

Ankley, GT, Niemi, GJ, Lodge, KB, Harris, HJ, Beaver, DL, Tillitt, DE, Schwartz, TR, Giesy, JP, Jones, PD and Hagley, C (1993) Uptake of planar polychlorinated biphenyls and 2,3,7,8-substituted polychlorinated dibenzofurans and dibenzo-p-dioxins by birds nesting in the Lower Fox River and Green Bay, Wisconsin, USA. Arch Environ Contam Toxicol 24:332–344.

Ankley, GT, Kuehl, DW, Kahl, MD, Jensen, KM, Butterworth, BC and Nichols, JW (2004) Partial life-cycle toxicity and bioconcentration modeling of perfluorooctane sulfonate in the northern leopard frog (*Rana pipiens*). Environ Toxicol Chem 23:2745–2755.

Barron, MG, Hansen, JA, and Lipton, J (2002) Association between contaminant tissue residues and effects in aquatic organisms. Rev Environ Contam Toxicol 173:1–37.

Boudreau, TM, Sibley, PK, Mabury, SA, Muir, DCG, and Solomon, K (2003a) Laboratory evaluation of the toxicity of perfluorooctane sulfonate (PFOS) on *Selenastrum capricornutum, Chlorella vulgaris, Lemna gibba, Daphnia magna,* and *Daphnia pulicaria.* Arch Environ Contam Toxicol 44:307–313.

Boudreau, TM, Wilson, CJ, Cheong, WJ, Sibley, PK, Mabury, SA, Muir, DCG, and Solomon, KR (2003b) Response of the zooplankton community and environmental fate of perfluorooctane sulfonic acid in aquatic microcosms. Environ Toxicol Chem 22:2739–2745.

Bowerman, WW, Best, DA, Grubb, TG, Zimmerman, GM, and Giesy, JP (1998) Trends of contaminants and effects in bald eagles of the Great Lakes Basin. Environ Monit Assess 53:197–212.

Brignole, AJ, Porch, JR, Krueger, HO, and Van Hoven, RL (2003) PFOS: a toxicity test to determine the effects of the test substance on seedling emergence of seven species of plants. Toxicity to Terrestrial Plants. EPA Docket AR226-1369. Wildlife International, Ltd., Easton, MD.

Chapman, PM, Caldwell, RS, and Chapman, PF (1996) A warning: NOECs are inappropriate for regulatory use. Environ Toxicol Chem 15:77–79.

Chapman, PM, Fairbrother, A, and Brown, D (1998) Critical evaluation of safety (uncertainty) factors for ecological risk assessment. Environ Toxicol Chem 17:99–108.

Desjardins, D, Sutherland, CA, Van Hoven, RL, and Krueger, HO (2001a) PFOS: A 96-hr toxicity test with the freshwater alga (*Anabaena flos-aquae*). Project Number 454A-110B. EPA Docket AR226-0186. Wildlife International, Ltd., Easton, MD.

Desjardins, D, Sutherland, CA, Van Hoven, RL, and Krueger, HO (2001b) PFOS: A 96-hr toxicity test with the marine diatom (*Skeletonema costatum*). Project Number 454A-113A. EPA Docket AR226-1030a056. Wildlife International, Ltd., Easton, MD.

Desjardins, D, Sutherland, CA, Van Hoven, RL, and Krueger, HO (2001c) PFOS: A 7-d toxicity test with duckweed (*Lemna gibba* G3). Project Number 454-111. EPA Docket AR226-1030a054. Wildlife International, Ltd., Easton, MD.

Dixon, DA (2001) Fluorochemical Decomposition Process. Theory, Modeling, and Simulation. W.R. Wiley Environmental Molecular Sciences Laboratory, Pacific Northwest National Laboratory, Richland, WA.

Di Toro, DM, McGarth, JA, and Hansen, DJ (2000) Technical basis for narcotic chemicals and polycyclic aromatic hydrocarbon criteria. I. Water and tissue. Environ Toxicol Chem 19:1951–1970.

Drottar, KR, and Krueger, HO (2000a) PFOS: A 96-hr toxicity test with the freshwater alga (Selenastrum capricornutum). Project Number 454A-103A. EPA Docket AR226-0085. Wildlife International, Ltd., Easton, MD.

Drottar, KR, and Krueger, HO (2000b) PFOS: A 48-hr static acute toxicity test with the cladoceran (*Daphnia magna*). Project No. 454A-104. EPA Docket AR226-0087. Wildlife International, Ltd., Easton, MD.

Drottar, KR, and Krueger, HO (2000c) PFOS: A 96-hr static acute toxicity test with the freshwater mussel (*Unio complamatus*). Project No. 454A-105. EPA Docket AR226-0091. Wildlife International, Ltd., Easton, MD.

Drottar, KR, and Krueger, HO (2000d) PFOS: A 96-hr static acute toxicity test with the saltwater mysid (*Mysidopsis bahia*). Project No. 454A-101. EPA Docket AR226-0095. Wildlife International, Ltd., Easton, MD.

Drottar, KR, and Krueger, HO (2000e) PFOS: A 96-hr shell deposition test with the eastern oyster (*Crassostrea virginica*). Project No. 454A-106. EPA Docket AR226-0089. Wildlife International, Ltd., Easton, MD.

Drottar, KR, and Krueger, HO (2000f) PFOS: A semi-static life-cycle toxicity test with the cladoceran (*Daphnia magna*). Project No. 454A-109. EPA Docket AR226-0099. Wildlife International Ltd., Easton, MD.

Drottar, KR, and Krueger, HO (2000g) PFOS: A flow through life-cycle toxicity test with the saltwater mysid (*Mysidopsis bahia*). Project No. 454A-107. EPA Docket AR226-0101. Wildlife International, Ltd., Easton, MD.

Drottar, KR, and Krueger, HO (2000h) PFOS: A 96-hr static acute toxicity test with the fathead minnow (*Pimephales promelas*). Project No. 454-102. EPA Docket AR226-0083. Wildlife International, Ltd., Easton, MD.

Drottar, KR, and Krueger, HO (2000i) PFOS: An early life-stage toxicity test with the fathead minnow (*Pimephales promelas*). Project No. 454-108. EPA Docket AR226-0097. Wildlife International, Ltd., Easton, MD.

Drottar, KR, Van Hoven, RL, and Krueger, HO (2001) Perfluorooctanesulfonate, potassium salt (PFOS): A flow-through bioconcentration test with the bluegill (*Lepomis macrochirus*). Project Number 454A-134. EPA Docket AR226-1030a042. Wildlife International, Ltd., Easton, MD.

Drouillard, KG, Fernie, KJ, Smits, JE, Bortolotti, GR, Bird, DM, and Norstrom, RJ (2001) Bioaccumulation and toxicokinetics of 42 polychlorinated biphenyl congeners in American Kestrels (*Falco sparverius*). Environ Toxicol Chem 20:2514–2522.

Duke, LD, and Taggart, M (2000) Uncertainty factors in screening ecological risk assessments. Environ Toxicol Chem 19:1668–1680.

Fisk, AT, Norstrom, RJ, Cymbalisty, CD, and Muir DCG (1998) Dietary accumulation and depuration of hydrophobic organochlorines: bioaccumulation parameters and their relationship with the octanol/water partition coefficient. Environ Toxicol Chem 17:951–961.

Gallagher, SP, van Hoven, RL, Beavers, JB, and Jaber M (2003a) PFOS: A reproduction study with Northern Bobwhite. Final Report. Project No. 454-108. USEPA Administrative Record AR-226-1831. Wildlife International, Ltd., Easton, MD.

Gallagher, SP, van Hoven, RL, Beavers, JB, and Jaber, M (2003b) PFOS: A reproduction study with mallards. Final Report. Project No. 454-109. USEPA Administrative Record AR-226-1836. Wildlife International, Ltd., Easton, MD.

Gallagher, SP, Casey, CS, Beavers, JB, and Van Hoven, RL (2004a) PFOS: A dietary LC$_{50}$ study with the mallard. Amended Report. Project No. 454-102. Available at EPA Docket AR-226-1735. Wildlife International Ltd., Easton, MD.

Gallagher, SP, Casey, CS, Beavers, JB, and Van Hoven, RL (2004b). PFOS: A dietary LC$_{50}$ study with the northern bobwhite. Amended report. Project No. 454-103. Available at EPA Docket AR-226-1825. Wildlife International Ltd.

Giesy, JP, and Kannan, K (1998) Dioxin-like and non-dioxin-like toxic effects of polychlorinated biphenyls (PCBs): Implications for risk assessment. Crit Rev Toxicol 28:511–569.

Giesy, JP, and Kannan, K (2001) Global distribution of perfluorooctane sulfonate in wildlife. Environ Sci Technol 35:1339–1342.

Giesy, JP, and Kannan, K (2002) Perfluorochemical surfactants in the environment. Environ Sci Technol 36:146A–152A.

Giesy, JP, and Newsted, JL (2001) Selected fluorochemicals in the Decatur, Alabama area. 3M Report. Project Number 178041. EPA Docket AR226-1030a161.

Giesy, JP, Ludwig, JP, and Tillitt, DE (1994) Dioxins, dibenzofurans, PCBs and colonial fish-eating water birds. In: Schecter, A (ed) Dioxin and Health. Plenum Press, New York, pp 249–307.

Giesy, JP, Dodson, S, and Solomon, KR (2000) Ecotoxicologial risk assessment for Roundup herbicide. Rev Environ Contam Toxicol 167:35–120.

Gledhill, WE, and Markley, BJ (2000a) Microbial metabolism (biodegradation) studies of perfluorooctane sulfonate (PFOS). I. Activated sludge/sediment. Lab ID Number 290.6120. EPA Docket AR226-1030a034. Springborn Laboratories, Inc., Wareham, MA.

Gledhill, WE, and Markley, BJ (2000b) Microbial metabolism (biodegradation) studies of perfluorooctane sulfonate (PFOS). II. Aerobic soil biodegradation. Lab ID Number 290.6120. EPA Docket AR226-1030a035. Springborn Laboratories, Inc., Wareham, MA.

Gledhill, WE, and Markley, BJ (2000c) Microbial metabolism (biodegradation) studies of perfluorooctane sulfonate (PFOS). III. Anaerobic sludge biodegradation. Lab ID Number 290.6120. EPA Docket AR226-1030a036. Springborn Laboratories, Inc., Wareham, MA.

Hanson, ML, Sibley, PK, Brain, RA, Mabury, SA, and Solomon, KR (2005) Microcosm evaluation of the toxicity and risk to aquatic macrophytes from perfluorooctane sulfonic acid. Arch Environ Contam Toxicol 48:329–337.

Jacobs RL, and Nixon, WB (1999) Determination of the melting point/melting range of PFOS. Project No. 454C-106. EPA Docket AR226-0045. Wildlife International, Ltd., Easton, MD.

Johnson, JD, Gibson, SJ, and Ober, RE (1984) Cholestyramine-enhanced fecal elimination of carbon-14 in rats after administration of ammonium [14-C] perfluorooctanoate and potassium [14-C] perfluorooctane sulfonate. Fundam Appl Toxicol 4:972–976.

Jones, PD, Hu, W, DeCoen, W, Newsted, JL, and Giesy, JP (2003) Binding of perfluorinated fatty acids to serum protein. Environ Toxicol Chem 22:2639–2649.

Kannan, K, Koistinen, J, Beckmen, K, Evans, T, Gorzelany, JF, Hansen, KJ, Jones, PD, Helle, E, Nyman, M, and Giesy, JP (2001a) Accumulation of perfluorooctane sulfonate in marine mammals. Environ Sci Technol 35:1593–1598.

Kannan, K, Franson, JC, Bowerman, WW, Hansen, KJ, Jones, PD, and Giesy, JP (2001b) Perfluorooctane sulfonate in fish-eating water birds including bald eagles and albatrosses. Environ Sci Technol 35:3065–3070.

Kurume Laboratory (2001) Final report. Bioconcentration test of 2-perfluoroalkyl (C = 4–14) ethanol [This test was performed using 2-(perfluorooctyl)-ethanol (Test substance number K-1518)] in carp. EPA Docket AR226-1276. Kurume Laboratory, Chemicals Evaluation and Research Institute, Japan.

Kurume Laboratory (2002) Final report. Biodegradation test of salt (Na, K, Li) of perfluoroalkyl (C = 4–12) sulfonic acid, test substance number K-1520 (test number 21520). Kurume Laboratory, Chemicals Evaluation and Research Institute, Japan.

Lange, CC (2001) The 35-d aerobic biodegradation study of PFOS. 3M Project Number E01-0444. EPA Docket AR226-1030a040. Pace Analytical Services, Minneapolis, MN.

MacDonald, MM, Warne, AL, Stock, NL, Mabury, SA, Solomon, KR, and Sibley, PK (2004) Toxicity of perfluorooctane sulfonic acid and perfluorooctanoic acid to *Chironomus tentans*. Environ Toxicol Chem 23:2116–2123.

Martin, JW, Mabury, SA, Solomon, KR, and Muir, DCG (2003a) Dietary accumulation of perfluorinated acids in juvenile rainbow trout (*Oncorhynchus mykiss*). Environ Toxicol Chem 22:189–195.

Martin, JW, Mabury, SA, Solomon, KR, and Muir, DCG (2003b) Bioconcentration and tissue distribution of perfluorinated acids in rainbow (*Oncorhynchus mykiss*). Environ Toxicol Chem 22:196–204.

Mayer, FL, Mayer, KS, and Ellersiek, MR (1986) Relation of survival to other endpoints in chronic toxicity tests with fish. Environ Toxicol Chem 5:737–748.

McCarty, LS, and Mackay, D (1993) Enhancing ecotoxicological modeling and assessment. Environ Sci Technol 27:1719–1728.

Meesters, RJ, and Schroder, HF (2004) Perfluorooctane sulfonate – a quite mobile anionic anthropogenic surfactant, ubiquitously found in the environment. Water Sci Technol 50(5):235–242.

Moody, CA, Martin, JW, Kwan, WC, Muir, DCG, and Mabury, SA (2002) Monitoring perfluorinated surfactants in biota and surface water samples following an accidental release of fire-fighting foam into Etobicoke Creek. Environ Sci Technol 36:545–551.

Newsted, JL, Beach, SA, Gallagher, SA, and Giesy, JP (2005a). Pharmacokinetics and acute lethality of perfluorooctane sulfonated (PFOS) to mallard and northern bobwhite. Arch Environ Contam Toxicol (Accepted, in press).

Newsted, JL, Coady, KK, Beach, SA, Gallagher, S, and Giesy, JP (2005b). Effects of perfluorooctane sulfonate on mallard (*Anas platyrhynchos*) and bobwhite quail (*Colinus virginianus*) when chronically exposed via the diet. Environ Toxicol Pharmacol (Accepted, in press).

Newsted, JL, Jones, PD, Coady, KK, and Giesy, JP (2005c) Avian toxicity reference values (TRVs) for perfluorooctane sulfonate (PFOS). Environ Sci Technol Accepted Online: DOI 10.1021/es050989.

Oakes, KD, Sibley, PK, Martin, JW, MacLean, DD, Solomon, KR, Mabury, SA, and van der Kraak, GJ (2005) Short-term exposures of fish to perfluorooctane sulfonate: Acute effects on fatty acyl-CoA oxidase activity, oxidative stress, and circulating sex steroids. Environ Toxicol Chem 24:1172–1181.

Obourn, JD, Frame, SR, Bell, RH, Longnecker, DS, Elliott, GS, and Cook, J (1997) Mechanisms for the pancreatic oncogenic effects of the peroxisome proliferator wyeth-14,643. Toxicol Appl Pharmacol 145:425–436.

OECD (Organization for Economic Cooperation and Development) (2002) Hazard assessment of perfluorooctane sulfonate (PFOS) and its salts. ENV/JM/RD (2002) 17/Final. 21 November, 2002. OECD, Paris.

Palmer, SJ, and Krueger, HO (2001) PFOS: A frog embryo teratogenesis assay-*Xenopus* (FETAX). Project No. 454A-116. EPA Docket AR226-1030a057. Wildlife International, Ltd., Easton, MD.

Palmer, SJ, Van Hoven, RL, and Krueger, HO (2002a) Perfluorooctanesulfonate, potassium salt (PFOS): A 96-hr static acute toxicity test with the rainbow trout (*Oncorhynchus mykiss*). Report No. 454A-145. EPA Docket AR226-1030a044. Wildlife International Ltd., Easton, MD.

Palmer, SJ, van Hoven, RL, and Krueger, HO (2002b) Perfluorooctanesulfonate, potassium salt (PFOS): A 96-hr static renewal acute toxicity test with the sheepshead minnow (*Cyprinodon variegatus*). Report No. 454A-146A. Wildlife International Ltd., Easton, MD.

Robertson, JC (1986) Potential for environmental impact of AFA-6 surfactant. EPA Docket AR226-1030a043. Beak Consultants Ltd., Missassauga, Ontario, Canada.

Rosenstrauch, A, Degen, AA, and Friedlander, M (1994) Spermatozoa retention by Sertoli cells during the decline in fertility in aging roosters. Biol Reprod 50:129–136.

Sanderson, H, Boudreau, TR, Mabury, SA, Cheong, WJ, and Solomon, KR (2002) Ecological impact and environmental fate of perflurooctane sulfonate on the zooplankton community in indoor microcosms. Environ Toxicol Chem 21:1490–1496.

Sargent, J, and Seffl, R (1970) Properties of perfluorinated liquids. Fed Proc 29:1699–1703.

Schaefer, EL, and Flaggs, RS (2000) PFOS: An activated sludge respiration inhibition test. Project Number 454E-101. EPA Docket AR226-0093. Wildlife International, Ltd., Easton, MD.

Schroder, HF (2003) Determination of fluorinated surfactants and their metabolites in sewage sludge samples by liquid chromatography with mass spectrometry and tandem mass spectrometry after pressurized liquid extraction and separation on fluorine-modified reverse-phase sorbents. J Chromatogr A 1020:131–151.

Sindermann, AB, Porch, JR, Krueger, HO, and Van Hoven, RL (2002) PFOS: an acute toxicity study with the earthworm in an artificial soil substrate. Project No. 454-111. EPA Docket AR226-1106. Wildlife International Ltd., Easton, MD.

Sohlenius, AK, Andersson, K, Bergstrand, A, Spydevold, O, and De Pierre, JW (1994) Effects of perfluorooctanoic acid, a potent peroxisome proliferator in rat, on morris hepatoma 7800C1 cells, a rat cell line. Biochem Biophys 1213:63–74.

Stephan, CE, Mount, DI, Hansen, DJ, Gentile, JH, Chapman, GA, and Brungs, WA (1985) Guidelines for deriving numeric National Water Quality Criteria for the protection of aquatic organisms and their uses. PB85-227049. U.S. Environmental Protection Agency, Washington, DC.

Suter, GW (1996) Toxicological benchmarks for screening contaminants of potential concern for effects on freshwater biota. Environ Toxicol Chem 15:1232–1241.

Sutherland, CA, and Krueger, HO (2001) PFOS: A 96-hr toxicity test with the freshwater diatom (*Navicula pelliculosa*). Project Number 454A-112. EPA Docket AR226-1030a055. Wildlife International, Ltd., Easton, MD.

Taniyasu, S, Kannan, K, Horii, Y, Hanari, N, and Yamashita, N (2003) A survey of perfluorooctane sulfonate and related perfluorinated organic compounds in water, fish, birds, and humans from Japan. Environ Sci Technol 37:2634–2639.

Tolls, J, and Sijm, DJHM (1995) A preliminary evaluation of the relationship between bioconcentration and hydrophobicity for surfactants. Environ Toxicol Chem 14:1675–1685.

Tolls, J, Haller, M, Graaf, ID, Thijssen, MATC, and Sijm, DTHM (1997) Bioconcentration of LAS: Experimental determination and extrapolation to environmental mixtures. Environ Sci Technol 31:3426–3431.

Urban, DJ, and Cook, NJ. (1986) Hazard Evaluation Division. Standard Evaluation Procedure: Ecological Risk Assessment. PB86-247657. U.S. Environmental Protection Agency, Arlington, VA.

U.S. Environmental Protection Agency (USEPA) (1995) Final Water Quality Guidance for the Great Lakes System: Final Rule. Fed Reg 60:15366–15425. USEPA, Washington, DC, USA.

U.S. Environmental Protection Agency (USEPA) (1997) Ecological Risk Assessment Guidance for Superfund: Process for Designing and Conducting Ecological Risk Assessments. Interim Final. EPA 540-R-97-006. Environmental Response Team, Edison, NJ.

U.S. Environmental Protection Agency (USEPA) (2001) Perfluorooctyl sulfonates; proposed significant new use rule. Fed Reg 65:62319–62333.

Van Hoven, RL, and Nixon, WB (2000) Determination of the water solubility of PFOS by the shake flask method. Project 454C-107. EPA Docket AR226-0052. Wildlife International Ltd., Easton, MD.

Van Hoven, RL, Stenzel, JI, and Nixon, WB (1999) Determination of the vapor pressure of PFOS using the spinning rotor gauge method. Project Number 454C-105. EPA Docket AR226-0048. Wildlife International, Ltd., Easton, MD.

Wildlife International (2000) Activated sludge respiration inhibition test. EPA Docket AR226-0092.

Wilkelski, M, Hau, M, Robinson, WD, and Wingfield, JC (2003) Reproductive seasonality of seven neotropical passerine species. Condor 105:683–695.

Wilkins, P (2001a) Perfluorooctanesulfonate, potassium salt (PFOS): An acute oral toxicity study with the honey bee. Study number HT5602. EPA Docket AR226-1017. Environmental Biology Group, Central Science Laboratory, Sand Hutton, York, UK.

Wilkins, P (2001b) Perfluorooctanesulfonate, potassium salt (PFOS): An acute contact toxicity study with the honey bee. Study number HT5601. EPA Docket AR226-1018. Environmental Biology Group, Central Science Laboratory, Sand Hutton, York, UK.

Yamada, T, and Taylor, PH (2003) Laboratory-scale thermal degradation of perfluoro-octanyl sulfonate and related substances. Environmental Sciences and Engineering Group, University of Dayton, Research Institute, Dayton, OH.

Manuscript received February 23; accepted June 22, 2005.

Index

175